THE HOC

The Wild Horses *of* Shackleford Banks *and vicinity*

THE HOOFPRINTS GUIDE TO

The Wild Horses *of* Shackleford Banks *and vicinity*

Written and Illustrated by

Bonnie U. Gruenberg

The Hoofprints Guide to the Wild Horses of Shackleford Banks and Vicinity

 The author's paintings, drawings, and photographs, including those used in this book, are available as posters and giftware at www.BonnieGPhoto.com

ISBN 13: 978-1-941700-16-7

Library of Congress Control Number: 2015952707

Published by Quagga Press, an imprint of Synclitic Media, LLC
1646 White Oak Road • Strasburg, PA 17579 • www.quaggapress.com

Also by the author

- *The Wild Horse Dilemma: Conflicts and Controversies of the Atlantic Coast Herds* (Quagga Press, 2015)
- *The Hoofprints Guide Series* (Quagga Press, 2015)
  Assateague • Chincoteague • Corolla
  Ocracoke • Shackleford Banks • Cumberland Island
- *Hoofprints in the Sand: Wild Horses of the Atlantic Coast*, Kindle Edition (Quagga Press, 2014)
- *Wild Horses of the Atlantic Coast: An Intimate Portrait*, Kindle Edition (Quagga Press, 2014)
- *Hoofprints in the Sand Wild Horses of the Atlantic Coast* (as Bonnie S. Urquhart; Eclipse Press, 2002)
- *Birth Emergency Skills Training: Manual for Out-of-hospital Providers* (Birth Guru/Birth Muse, 2008)
- *Essentials of Prehospital Maternity Care* (Prentice Hall, 2005)
- *The Midwife's Journal* (Birth Guru/Birth Muse, 2009)

Forthcoming

- *Wild Horse Vacations: Your Guide to the Atlantic Wild Horse Trail with Local Attractions and Amenities*(Quagga Press, 2015)
  Vol. 1: Assateague, MD | Chincoteague, VA | Corolla, NC
  Vol. 2: Ocracoke, NC | Shackleford Banks, NC | Cumberland Island, GA
- *Wild Horses! A Kids' Guide to the East Coast Herds* (Quagga Press, 2015)
- *Birth Emergency Skills Training*, 2nd Edition (Synclitic Press, 2015).

Introduction

The water was mirror of molten glass, reflecting distorted caricatures of the horses as they grazed. I had feared that the capricious weather would cut our trip short, but instead the dense clouds and errant sunbeams created a wonderland. When Nena and Woody Hancock generously offered to take me on a tour of the private Cedar Island wild horse refuge, I expected blazing heat, relentless sun, and voracious insects. I expected to feel like an intruder, disrupting the natural order with my presence.

Instead, I stood knee deep in quiet water under the shelter of leaden clouds, telephoto trained on the peaceful equine family that was foraging in the marsh. Protected within the lushness of their Eden, they exuded peace and belonging, and I felt an inexplicable sense of communion with the wild things and their watery world.

When I first started researching the wild horses in the mid-1990s, I was surprised to find that wild horses lived on a number of Atlantic barrier islands and had once ranged along much of the Atlantic coast. They made their first hoofprints there not long after the arrival of early European settlers, and in time they ran free on innumerable North American islands and peninsulas from the Caribbean to Canada. I learned that small herds remained on the coast of Virginia, North Carolina, Maryland, and Georgia; on Sable Island, off Nova Scotia, Canada; and on Great Abaco Island in the Bahamas. Each population of horses has its own character, its own history, and its own set of problems. In most cases, these animals have made a unique contribution to local history, and each herd has its own detractors and defenders.

After my first book, *Hoofprints in the Sand: Wild Horses of the Atlantic Coast,* was published in 2002 by Eclipse Press, I dove in deeper, interviewing experts, evaluating the evidence, and monitoring the herds. I explored management conflicts that encompassed political, economic, and cultural issues as well as purely scientific ones. I studied storms and shipwrecks, equine behavior and genetics, history, epidemiology, barrier-island dynamics, sea-level rise, beach development, and the perpetual clash of viewpoints. I studied hundreds of documents, from historical papers to scholarly journals to court transcripts, so that I might accurately present the pertinent issues. Distilling all this information, I tried to present all sides of the issues fairly so that readers might reach their own conclusions. The result is *The Wild Horse Dilemma: Conflicts and Controversies of the Atlantic Coast Herds* (Quagga Press, 2015) the most comprehensive work ever published about these horses.

Wild Horse Dilemma is exhaustively researched, copiously documented, and peer-reviewed; but at 600 pages it may be too long for many people eager to learn about a particular herd. For readers with limited time, I created the Hoofprints Series. Excerpted from *Wild Horse Dilemma* and containing additional photographs, each Hoofprints book presents a single Atlantic Coast herd in sufficient detail to satisfy both the layman and the academic.

I take all my wild-horse photographs though telephoto lenses that let me to keep my distance. When horses approached, I retreated. My goal has been to remain so peripheral to their lives, they will forget that I am nearby. Because countless people have stroked them, fed them, and lured them, some can be momentarily docile, occasionally indifferent, or routinely bold and pushy in the presence of people. As anyone bitten or trampled can attest, they are no less wild than horses that avoid human contact. When we impose ourselves and our desires on their lives, when we habituate them to our presence, when we teach them to approach us for food and attention, we rob them of their wildness. When we treat them as we would their domestic counterparts, we miss the opportunity to observe them in a natural state, that is, to appreciate the things that make them irresistibly attractive. We miss the very point of driving past thousands of their tame kin to seek them out. We create something like a petting zoo hazardous to us and to them. If we truly love and respect wild creatures, we must learn to stand back and enjoy watching them from afar. Only then can they—and we—know the real meaning of wildness.

As the earth's dominant species, we have the power to preserve or destroy the wildlife of the world and the ecosystems in which they live. The choices we make regarding wild horses are far-reaching. We alter their destiny whether we act or choose to do nothing. We can begin to deal wisely with wild horses by understanding the facts and discovering how the threads of their existence are woven into the tapestry of life. Only through understanding can we hope to make rational, educated decisions about the welfare of these fascinating, inspiring animals.

Bonnie U. Gruenberg
Strasburg, Pennsylvania
October 1, 2015

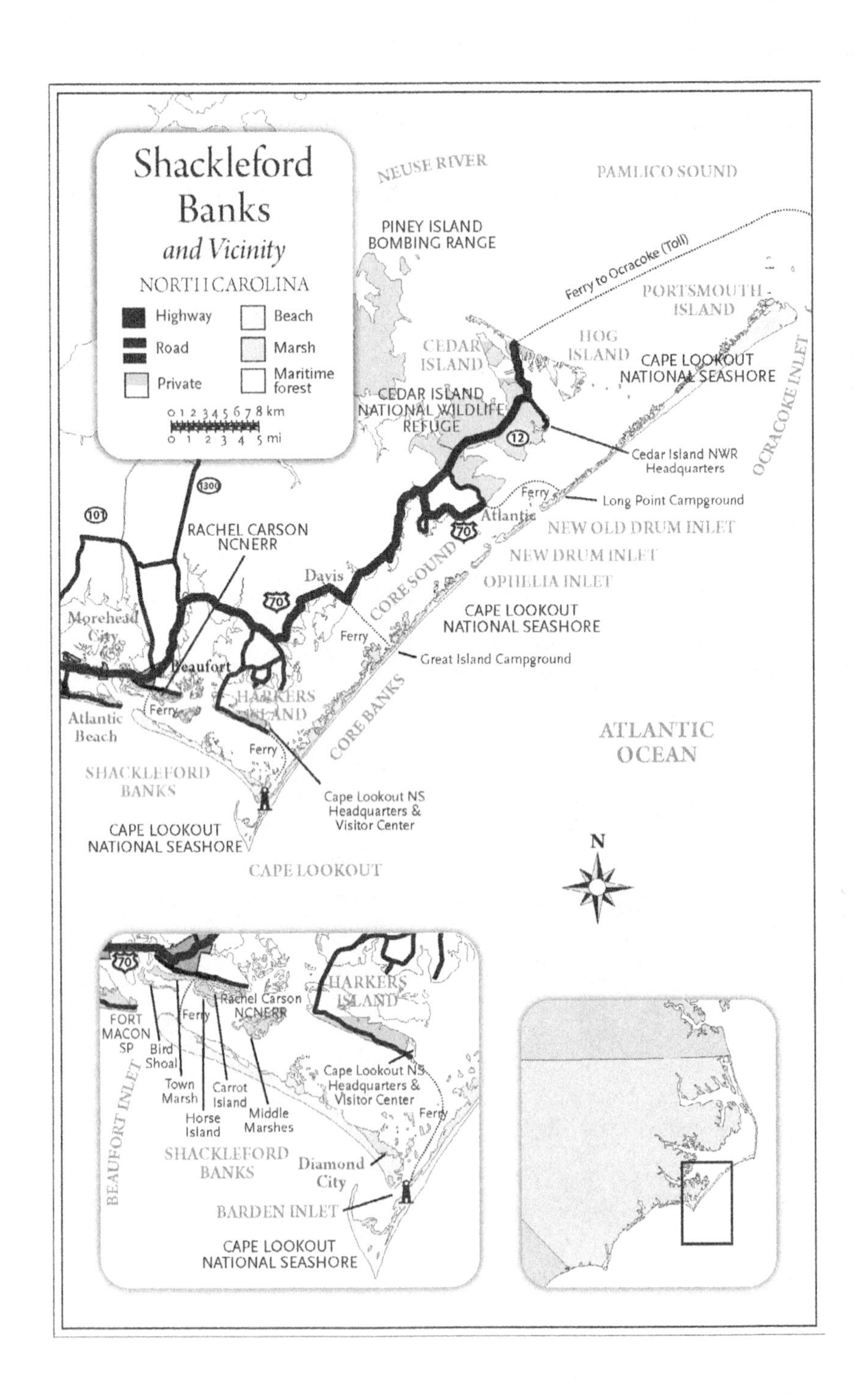

Shackleford Banks
and Vicinity
NORTH CAROLINA
Highway
Road
Private
Beach
Marsh
Maritime forest
0 1 2 3 4 5 6 7 8 km
0 1 2 3 4 5 mi
NEUSE RIVER
PAMLICO SOUND
PINEY ISLAND BOMBING RANGE
Ferry to Ocracoke (Toll)
PORTSMOUTH ISLAND
CEDAR ISLAND
HOG ISLAND
CAPE LOOKOUT NATIONAL SEASHORE
OCRACOKE INLET
CEDAR ISLAND NATIONAL WILDLIFE REFUGE
Cedar Island NWR Headquarters
Ferry
Long Point Campground
Atlantic
NEW OLD DRUM INLET
NEW DRUM INLET
OPHELIA INLET
RACHEL CARSON NCNERR
Davis
CORE SOUND
CAPE LOOKOUT NATIONAL SEASHORE
Morehead City
Beaufort
Great Island Campground
HARKERS ISLAND
CORE BANKS
Atlantic Beach
ATLANTIC OCEAN
SHACKLEFORD BANKS
Cape Lookout NS Headquarters & Visitor Center
CAPE LOOKOUT NATIONAL SEASHORE
CAPE LOOKOUT
N
FORT MACON SP
Bird Shoal
Town Marsh
Carrot Island
Horse Island
Middle Marshes
Rachel Carson NCNERR
BEAUFORT INLET
Diamond City
BARDEN INLET

The seal-brown mare known to the Cape Lookout National Seashore staff as 13R was thirsty. The day was warm, and the only available water came at the price of hard labor. When rainfall is sparse, the horses of Shackleford Banks dig holes in the sand, which fill with freshwater when they reach the water table. After grazing in the relentless sun for most of the morning, the band led by the rugged bay stallion 14K arrived at the drinking spot. The alpha mare, the 9-year-old sorrel 14L, quickened her step at the scent of water. As they rounded the dunes and stepped into the clearing, the stallion moved in front of the sorrel with animated steps, head and tail high, prepared to defend his mares from competing stallions. But the clearing was empty except for five crows that took to air, complaining in raspy voices.

As the dominant animal in the group, the muscular stallion drank first, drawing easily at the water that had accumulated in the depression. Next in line was the sorrel, distinctive for her shining red-gold coat and a milky blaze that coursed down her face and pooled over her left nostril. She took a few sips of standing water, then dug deeper to let more flow into the hollow.

The seal-brown mare's mouth felt as dry as the sand that blew around her fetlocks, but she was 4 years old and low in rank. Protocol demanded that she wait her turn.

Two-year-old 15U pushed around the sorrel to steal a sip out of turn, but the lead mare flattened her ears and threatened to bite. The filly retreated reluctantly. The past winter had left the adolescent looking somewhat ragged—an undershot jaw and angular limb deformity had made her life challenging, as did her position at the very bottom of the pecking order.

As the alpha mare signaled her satiation and stepped away, she was replaced by the filly's dam, 22-year-old Number 16, who was leaner and less muscular than the lead mare. Despite abundant forage, the mare remained thinner than other adults in the herd, probably because of her advanced age, repeated reproduction, and recent lactation. The elderly mare pawed vigorously to excavate the depression, then lowered her muzzle to sip while the brown mare and the filly stood by, obvious distress in their eyes.

When the seal-brown mare finally took her turn, she had barely lowered her lips to the murky fluid when the stallion decided that it was time to leave. He aggressively pushed his harem into motion with flattened ears and snaking head. The mares trotted beyond his reach and obediently ascended the dune, but when the stallion looked away, the brown doubled back and rushed toward the drinking hole. Apparently, the young mare had resolved to slake her thirst regardless of the consequences.

The stallion was incensed at her insolence. This was not the first time the mare had flouted his authority. He spun and flew at her with teeth bared. The brown mare nimbly evaded him, escaping to the safety of the dunes. Satisfied that he had driven her from the water source, the bay stallion shifted his attention to the other mares and pressed the herd north. The seal-brown saw opportunity and spun away from her herdmates in another unauthorized rush to the water hole. The stallion appeared dumbfounded—this low-ranking upstart was persisting in challenging his decree?

Ears flat, head low, and thoroughly enraged, the muscular bay flew at her like an arrow, and she poured on the speed, raising a cloud of fine sand. With the stallion interposed between the water hole and the band, the seal-brown had no choice but to rejoin the herd and allow the bay to drive her over the dune to the grasslands.

An hour later, a younger battle-scarred bay stallion with the freezebrand 2L was taking his turn at the water hole, accompanied by his two mares. They lifted their heads to the unexpected sound of galloping hoofbeats and saw the seal-brown mare round the dune at full tilt and rush straight to the water hole in total disregard of social protocol. Recklessly, the mare thrust her muzzle into the hole.

Her stallion arrived seconds later, infuriated. But as he followed her into the clearing, he found himself face to face with a surprised and indignant 2L. The older stallion hesitated for an instant, and then launched himself at 2L, catching him off guard. The younger stallion fell back, off balance, then rallied and descended on his rival, biting him savagely and raking him with his front hooves. The older bay, to this point brash and confident, suddenly realized that he was

The sorrel alpha mare, 14L (far left), admonishes 2-year-old 15U, an impatient adolescent, to wait her turn. The bay harem stallion looks on, thirst slaked. On the far right, the seal-brown mare 13R gazes longingly at the sorrel, the set of her ears communicating her discomfiture.

With ears flat and hind leg cocked, mare number 16, age 22, warns her filly to back off while she drinks. The filly scrambles away in a fearful, submissive posture and bumps into the seal-brown mare, who similarly warns her out of her personal space with a flat-eared head toss.

Finally it was the brown mare's turn, and she pushed the adolescent filly aside to claim her rightful place at the water hole.

The stallion decided that it was time to move on and drove his band from the water hole.

Head low and ears flat, the stallion drove the brown mare from the water and pressed her toward the other mares. The adolescent filly saw an opportunity and began sucking down the contents of the muddy pool.

But the brown mare was not finished drinking. She dodged the stallion and cantered back to the water hole.

Her insolence was not to be tolerated. The stallion flattened his ears and galloped after her, but she was faster than the stallion and nimbly kept out of reach.

outmatched and withdrew. He tucked his tail in submission and retreated across the dunes.

The seal-brown mare gulped greedily at the water hole, uninterested in the drama around her. When her mate finally returned, exhausted and humiliated, the little rogue mare was waiting with a self-satisfied expression, her thirst now slaked. She allowed him to herd her back to the band without a protest.

As the reader will note, the preceding tale refers to these horses only by color, number, age or other characteristics. It was a challenge

Eventually, the bay stallion succeeded in driving the mare from the water hole, but she refused to give him the respect his station demanded, and he was frustrated. Determined to exert his power over someone, he changed course and charged straight at the author, who was standing nearly 2 bus lengths away documenting the events through a telephoto lens. She clutched her camera and ran! Fortunately this was the desired response—the stallion turned and smugly returned to his mares.

The stallion drove his harem away from the water hole, but the seal-brown mare remained thirsty and determined. Sometime later, the author heard the hammering of hoofbeats and looked up to see the mare rushing back to the water hole with the bay stallion in tight pursuit.

to create an engaging account of this water-hole drama without using names, especially when two look-alike bay stallions squared off for battle. Moreover, all these horses do indeed have formal names, but the Cape Lookout NS asked that the author omit them from this book.

Each year, graduate students of Dr. Daniel Rubenstein, a professor of ecology and evolutionary biology at Princeton University who has been conducting research on Shackleford Banks since the 1970s, name the new foals of the Cape Lookout herd. They give each new filly or colt a name that begins with the first letter of its dam's name and conforms to an annual theme—Shakespeare, music, cartoon characters, countries, Biblical figures, and so on.

These names are used by the Foundation for Shackleford Horses on public Web sites, by reporters in newspaper articles, and by Carmine Prioli in his comprehensive work, *The Wild Horses of Shackleford Banks* (2007). They are even used among biological professionals at the Cape Lookout NS and in official documents accessible by the public (such as Bardenhagen, Rogers, & Borrelli, 2011). But in conversations with lay people, the agency now refers to horses by number only "to help the public see them as more wild" (S. Stuska, personal communication, February 19, 2013)—an unexpected stance for an agency with a long history of denying horses status as wildlife.

Surprisingly, choosing to name horses in a wild herd or publicize these names can become a point of contention. Names empower us to organize and understand a forbidding universe of impersonal information, but also affect our emotions and reason. Upon naming wild animals, some people feel a sense of ownership that creates a perceptual shift, which can cause them to mistake the natural attributes of the creature for those that they assign. Often it takes a determined effort to allow wild creatures to be their authentic selves without remaking them through our own projections.

Even naming inanimate objects creates emotional attachment. The city of Boston, Massachusetts, has successfully encouraged residents to take responsibility for keeping fire hydrants clear of snow by allowing businesses, organizations, and private citizens to adopt and name them.

Naming wild animals stimulates empathy. Some recreational boaters change their behavior around wild orcas in Puget Sound,

As they approached, the stallion abruptly found himself face to face with another alpha stallion using the water hole. They scuffled, and the bold bay quickly realized that he was outmatched.

The younger stallion 2L drove the harem stallion 14K into a humiliated retreat.

Washington, when they have learned the identities of individual whales. In one case, a private boater was speeding dangerously, pursuing whales and disregarding safety guidelines. The driver was surly and uncooperative until the naturalist explained that the whale in question was Granny, a 99-year-old pod leader. Milstein (2011, p. 11) wrote, "Granny's name was . . . easy for visitors to remember, and it enfolded her matriarchal position into a culturally comprehensible package, eliciting similarities with humans, in this case familial relations, and likely mediating points of connection and even empathy." Upon recognizing the whale as an individual, the boater willingly yielded space to the elderly Orca.

Similarly, at Assateague Island NS in Maryland, officials use horse names to engage and educate the public. When Dr. Ronald Keiper began his research with the herd in the 1970s, he assigned each band a letter, and each pony in the band a number. Each foal was labeled with its mother's identifier, plus a letter indicating its birth year—A for 1976, B for 1977. To the initiated, the birth year, mother, grandmother, and great-grandmother are immediately evident in the designation of foal N2B-E-H.

This system allows managers and researchers to follow the maternal pedigree of the entire herd easily and indefinitely, and to trace maternal lineage back for nearly four generations, without time-consuming use of reference lists. The designations alone can help a scientist keep entire genetic lines in mind as they surge and ebb in often surprising ways, as prolific lines stop reproducing and vanish while poorly represented lines unexpectedly increase.

However useful, these designations are not quite names. Though Keiper's work was strictly scientific, it is evident from his writing that he warmed to the animals he studied. Eventually he saw a need to use real, meaningful names. He wrote (1985, p. 19), "As in other behavioral studies, some ponies earned conventional names related to their appearance, personality, or family life." He named Comma for the white crescent on his forehead, Nasty for her aggressive behavior, and Park Service Pony for a marking resembling the agency's uniform patches. He named an elderly mare Irene because, like his grandmother of the same name, she produced three female offspring, followed by a male.

Today, most horses in the Assateague NS herd have names, and the seashore keeps an identification book in its visitor center to help people investigate horses or nominally adopt their favorites. The nonprofit Assateague Island Alliance maintains a Web site where visitors can identify and "adopt" these horses or bid on naming rights for new foals, with proceeds benefitting the seashore.

Mustangs in many of the Western herds have names. Wild horse advocates and researchers name Pryor Mountain horses, and even the U.S. Bureau of Land Management refers to these horses by name on its official Web site. The herd gained a high public profile after Ginger Kathrens featured them in three extraordinary documentaries, following the life of the wild stallion Cloud, for the Public Broadcasting System series *Nature.* Kathrens balances scientific objectivity with genuine concern and interest and paints honest, yet emotionally stirring portraits of mustangs and the challenges of their mountainous Montana range. These films, along with numerous books, articles, and plastic Breyer® horse models, have promoted understanding of wild horses in general by helping people appreciate them as individuals. Cloud lent his name to a nonprofit support group, the Cloud Foundation.

Raising awareness of these horses has not spoiled their wild character in their natural range, but it has transformed the hearts and minds of the public. Whereas the average 2012 adoption rate of BLM horses and burros was 31.37%, 100% of the gathered Pryor Mountain horses were quickly adopted or placed in private care—as in every previous year. The BLM infrequently offers Pryor Mountain horses for adoption, and few horses are gathered from the herd.

Most often, people choose names for animals when they feel affinity and numbers when they are, or wish to be, disengaged, as in the case of laboratory rabbits or feedlot steers. In Alberta, Canada, horses bred specifically for meat production and fattened on feedlots are somehow deemed more suitable for slaughter than companion horses, because "Purpose bred meat horses do not have names, nor have they served humans in another role" (Alberta Equine Welfare Group, 2008, p. 31). Before modern medicine increased the survival rate of human infants, parents often delayed naming their babies until it became clear that they would live.

In most of the Atlantic coast herds, managers struggle with the problem of visitors who approach and interact with wild horses as if they were already tame. As Dwyer (2007, p. 86) wrote, "we tend to generalize the successes we have with our pets to all animals. If we can have a loving relationship with a dog, why not one with a fox or a wolf? If we can get a horse to love us, why not a zebra?" Close contact with wild creatures, however, often ends badly, resulting in injuries to humans and bold, aggressive behavior in wildlife. In our culture, animals that hurt people are usually removed or destroyed.

At Cape Lookout NS, administrators say officially that numerical identifiers will emphasize the enigmatic aura of wildness that keeps visitors at a distance. Not everyone accepts this explanation, however. Years of controversy over drastic state and federal management practices, such as mass euthanasia of horses, pressing for unsustainably small herd size, and removing other livestock, have created suspicion on both sides of the public-private divide. Naming can lead to attachment. Whereas few may have noticed or cared about the sudden disappearance of some horse with a name like 12M, the death of a white foal named Spirit of Shackleford generated widespread sympathy and support for the herd. In this climate, some imply that the seashore numbers its horses to prevent public sentiment from complicating the next unpopular decision or simple mishap.

For centuries, free-roaming horses have lived and died on the 60-odd miles (97 km) of North Carolina islands that now constitute Cape Lookout NS. Now they remain only on 3,000-acre/1,200-ha Shackleford Banks, a narrow ribbon of sand roughly 9 mi/14.5 km long that runs perpendicular to the southernmost end of Core Banks.

North and South Core Banks extend for 44 mi/71 km from southwest to northeast, a long, slender stretch of low dunes, grasses, shrub thickets, maritime forest, and spreading salt marshes. Periodically, inlets divide the island, opening during storms and closing again in a matter of months or years. At this writing, three are open: the sporadically maintained New Drum Inlet; New Old Drum Inlet, reopened by Hurricane Irene in 2011; and Ophelia Inlet, opened by Hurricane Ophelia in 2005. The natural migration of the chain is unimpeded by human structures, and they are free to respond to the pressures of waves and overwash. From north to south, the landscape changes from wide tidal sand flats at Portsmouth Island to continuous dunes at Cape Lookout. Small freshwater marshes form in the depressions between dunes or where sandbars or spits close the mouth of a small bay. A jetty built in the early 1900s led to the formation of a sandy hook at the western end of Cape Lookout.

After hundreds of years of relative isolation on barrier islands, the North Carolina Banker Horses still resemble the Spanish Jennet that carried the conquistadors.

Compared to Core Banks, Shackleford displays more environmental diversity, with dunes as high as 44 ft/13 m at the western end, a dense maritime forest, marshes, and grassland. About 3.5 mi/5.6 km east of Beaufort Inlet, the island's profile shifts abruptly from high dunes to low overwash flats.

Here, as on Assateague, Currituck Banks, and Ocracoke, many believe that the original horses swam ashore from Spanish shipwrecks long before the English colonized the New World. The shoals off Cape Lookout and along the banks are certainly treacherous. One of the region's earliest cartographers named Cape Lookout *Promontorium tremendum*, or "horrible headland." Ships, mostly bound to and from the English colonies, often passed this hazard carrying horses and other livestock, especially in the mid-to-late 1600s and 1700s. Again, shipwrecks might have brought horses to the island, but there is no proof, and perhaps no possibility of proof.

Sue Stuska, the Park Service biologist who manages the Shackleford herd, says that she does not know of any convincing evidence that the original horses arrived by shipwreck:

Several horses in the Shackleford Banks herd carry the rare Q-ac blood type variant, which strongly suggests Colonial Spanish ancestry.

> As a scientist, I can't say that they swam to the island from sinking ships. The horses could have come by way of early explorers, colonists, or overland trade. Any shipwrecked horses that reached the island would have been the same type of horses that the colonists already had. The islands were like a common grazing ground. Anyone could take an animal over and drop him off. (S. Stuska, personal communication, May 26, 2010)

Few records exist, and the horses have been there for so long, many generations grew up believing that they were native to the islands. Edmund Ruffin, an agriculture authority who visited the area around 1858 and published his observations three years later, concluded that these horses were at that time a genetically unique, closed population, and that outside horses had not contributed their genes to the herd in recent history:

> The race, of course, was originally derived from a superior kind or breed of stock; but long acclimation, and subjection for many generations to this peculiar mode of living, has fixed on the breed the peculiar characteristics of form, size, and

The Q-ac blood type variant is also present in the Pryor Mountain mustang herd of Wyoming and Montana.

> qualities, which distinguish the "banks' ponies." It is thought that the present stock has suffered deterioration by the long continued breeding without change of blood. (1861, pp. 132–133)

Ruffin went on to say that introducing the blood of other breeds to the herd might decrease their ability to withstand the harsh environment of the Banks and that mainland horses "if turned loose here, would scarcely live through either the plague of bloodsucking insects of the first summer, or the severe privations of the first winter" (1861, p. 133).

Blood samples analyzed in the late 1980s showed that Shackleford horses were genetically similar to Ocracoke horses, "from which breeding stock reportedly was derived," but have other genes in common with draft breeds (Goodloe, Warren, Cothran, Bratton, & Trembicki, 1991, p. 417). This study concluded further "the Cumberland, Ocracoke, and Assateague horse populations are not genetically unique" (p. 418) in relation to the blood markers studied and that the horses did not qualify for retention on federal lands. The authors reserved judgment regarding the Shackleford herd due to small sample size: two of four horses died when the researchers immobilized them for phlebotomy, and the team halted sampling.

The findings of Goodloe et al. (1991) correspond with the historical evidence. Throughout the 20th century and probably earlier, the herds at Cumberland, Ocracoke, and Assateague Islands saw numerous infusions of outside bloodlines, mostly from American breeds. The Shackleford and Corolla herds are unrelated members of the same rare breed, probably derived from the same foundation population, but isolated from each other for hundreds of years. As discussed previously, the horses of North Carolina's Outer Banks probably descend from Spanish Horses from the West Indies, Chickasaws from the mainland, and perhaps Narragansett Pacers, racing horses, and saddle horses from New England and elsewhere. The Shackleford herd probably saw few introductions in the 20th century aside from genetically similar horses from nearby islands and the mainland.

On the other hand, Conant, Juras, and Cothran (2012, p. 60), write that

> The Marsh Tackies, Florida Crackers and Shackleford Banks populations show similarity to each other . . . and all show affinity to the Puerto Rican Paso Fino and other South American Iberian breeds, which is consistent with popular history and the Colonial Spanish designation for these populations. Given that early European explorers and settlers often obtained livestock from Puerto Rico and other Caribbean islands before heading to the eastern coast of the United States, this signature may truly show shared ancestry, which is rather remarkable after 400 years of separation.

Certain DNA markers are very characteristic of Spanish bloodlines, and these occur at relatively high frequency within the population. Sponenberg (2011) writes, "Due to the inheritance pattern of these markers it is easily possible for an absolutely pure Colonial Spanish Horse to have missed inheriting any of the Iberian markers. It is likewise possible for a crossbred horse to have inherited several." DNA typing provides useful insights and information, but leaves much open to interpretation. Sponenberg (2011) says,

> These techniques do offer great help in verifying the initial results of historic and phenotypic analysis, but are by themselves insufficient to arrive at a final conclusion. . . .
>
> A conservation program based heavily on blood types without considering other factors could then easily exclude

Racing Marsh Tackies on the beach was once a winter tradition, following the fall harvest. Carolina Marsh Tackies persist as a rare, historic breed. In 2012, 17 raced before a crowd of more than 7,000 spectators on Coligny Beach, Hilton Head, SC. Photograph by Anthony Surbeck via Wikimedia Commons.

> the very horses whose conservation is important, and could include some that should have been excluded. Therefore, conformational type is a more important factor than blood type or DNA type, and will always remain so. It is impossible to determine the relative percentage of Spanish breeding in a horse through blood typing or DNA typing, at least currently.

Several variants (including Pi-W and Q-ac) clearly denote old Spanish origin. The only other breeds known to carry the Q-ac variant are the Puerto Rican Paso Fino and the Pryor Mountain wild horses of Wyoming and Montana.

Cothran says that although the marker cannot prove where or how this ancestry came about, he believes that the North Carolina Banker Horses have descended from a very old lineage of Spanish horses:

> The Q-ac is a very uncommon variant that we see in what is called the Q blood group. Horses have blood groups just like people do. When I was doing some of my early work on mustangs, I was noticing this uncommon variant in some populations. And when I say uncommon, I mean that in a herd where it is relatively common, you'll see it in 2–5% of the population. When I see that variant in a population, I consider that to be evidence of old Spanish bloodlines. . . .
> I observed the highest frequency in the Puerto Rican Paso Fino, which is a very old Spanish breed. That population had a frequency of that variant at 5 percent—the highest

> of anywhere. I did not see the variant in any of the North American breeds such as the Morgan, the Saddlebred or the Quarter Horse. . . . On Shackleford Banks, only 3 or 4 horses in the herd had the Q-ac variant. At the population level, it is saying something. It is likely that there is old Spanish blood in that population . . . They show a high similarity to some of the English pony breeds like the Dales. (Personal communication, April 20, 2011)

Shackleford horses and certain Western herds of Spanish mustangs contain individuals with the rare, archaic Q-ac blood type variant. The most likely explanation is that both Western and Eastern Spanish horses descend from the same source population—Spanish ranches established in the 1500s in the Caribbean. It is well documented that Native American traders used horses from Western populations of pure Spanish horses to transport goods and sold the animals to lowland farmers in the Carolinas.

Another possible, but less likely, explanation relates to the use of horses to patrol the American coast during WWII. In June 1942, a U-boat landed four German saboteurs equipped with explosives, ammunition, cash, and forged documents on a beach near Amagansett, Long Island, New York. Another contingent landed on Ponte Vedra Beach, near Jacksonville, Florida. Federal authorities quickly caught them, but the episode pointed out the weakness of coastal defenses. Within months, the Coast Guard created an armed Beach Patrol with motorized, canine, and mounted elements to cover the Atlantic, Pacific, and Gulf coasts. Mounted units eventually served everywhere in the East except parts of New England. The U.S. Army Remount Service provided saddles and other gear. Because it also had procurement depots in Wyoming, Montana, and Nebraska, it provided more than 3,200 horses, some of them unbroken mustangs. There are no accessible records of where each horse originated; where each was sent; or what percentage were mares, stallions, or geldings. After the war, the horses were no longer needed and expensive to maintain, and most were immediately removed from service. Few records reveal the fate of these horses. Auctions in Ocean City, Maryland, Virginia Beach, Virginia, and elsewhere that began in 1944 do not account for all the horses. The Coast Guard seems to have returned many to the Army, but might have abandoned some of them

Members of the U.S. Coast Guard Beach Patrol at Hilton Head Training Station, South Carolina ("Beach Pounders," *circa* 1943). Mounted Guardsmen patrolled the beaches along the Atlantic coast south of New England during World War II. Many of their horses, provided by the U.S. Army Remount Service, were mustangs from the West. Did their genes enter the barrier island herds? Photograph courtesy of the U.S. Coast Guard Historian's Office.

or sold them to coastal stockmen who added them to their free-roaming herds. Could horses that carried the Q-ac allele—military horses of old Spanish lineage or Beach Patrol horses from the Pryor Mountains or elsewhere—have added their genes to the Shackleford herd? There is no hard evidence that they did, and there may be no way to find out; but it remains an interesting possibility.

The first known residents of these islands were the Coree Indians, who had lived on and around them seasonally since prehistoric times. In fact, the name Core Banks may be derived from *Coree*. A settlement was established in the Core Sound area by 1710. The Tuscarora and allied tribes staged a coordinated attack on the colony in 1711, igniting years of violence before a peace treaty ended the Tuscarora War in 1715. Subsequently, Eastern North Carolina opened to further

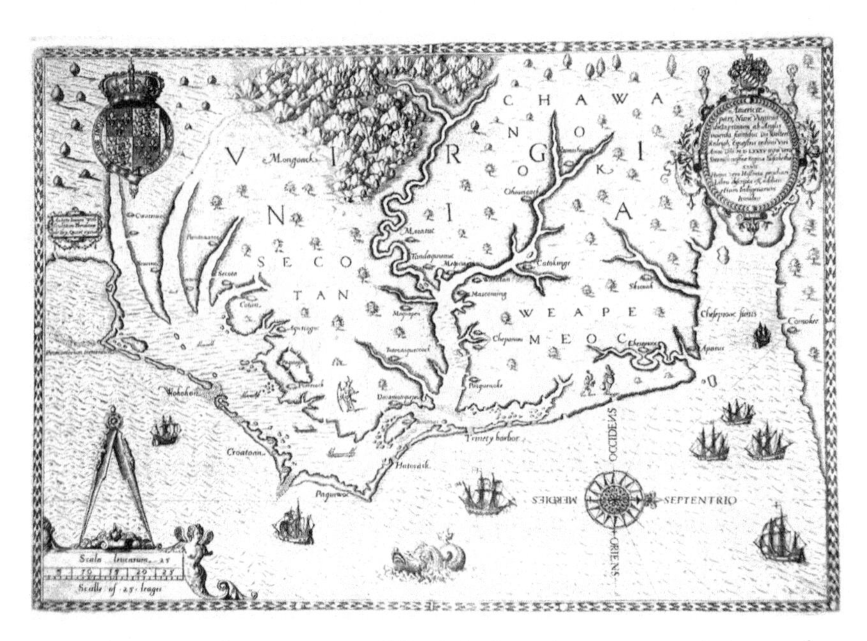

Theodor De Bry's 1590 map of Virginia, based on manuscript maps by John White, was the first accurate representation of the Outer Banks and vicinity. Courtesy of the Library of Congress, Geography and Map Division.

colonization after many Tuscarora migrated to western New York, and other native peoples succumbed to disease or were displaced onto reservations.

In 1713 John Porter acquired Core Banks and Shackleford Banks, then parts of a single island connected by a tidal flat known as "The Drain." Porter sold the 7,000-acre/2,800-ha island to Enoch Ward and John Shackleford. When they divided the properly in 1723, Shackleford took the western part and gave it its current name.

Colonists of predominantly English descent settled the island in the 1760s, and by the mid-1800s, Shackleford Banks was home to more than 600 people in several communities who called themselves "Ca'e Bankers" (dialect for "Cape Bankers"). The U.S. Coast and Geodetic Survey topographic map of 1853 shows about 20 residences along the sound side of Lookout Woods in the area that became known as Diamond City. Although most of the Cape Bankers lived in this area, there were homes all along the sound side of the Banks.

Diamond City, named after the distinctive diamond pattern of the nearby Cape Lookout Lighthouse, was the largest town ever

Diamond City derived its name from the distinctive diamond pattern of the nearby Cape Lookout Lighthouse. Today wild horses roam where a village once prospered.

established on Shackleford Banks. It was situated on the east end before Barden Inlet divided the island from Core Banks, and it boasted 500 residents. People had lived on the east end of Shackleford since the Banks were settled, but the community was not named Diamond City until around 1885.

Other Shackleford villages included Wade's Shore, Mullet Pond, Bell's Island, and Windsor's Lump. The island supported an oyster house, a factory that extracted oil from porpoises, a crab-packing plant, schools, businesses, and churches. Feral sheep, goats, hogs, cattle, and horses freely roamed the island, while the Bankers lived in fenced homesteads built from maritime-forest timber and salvage from shipwrecks.

A Reconstruction-era report of the North Carolina Geological Survey noted,

> These islands are inhabited by a hardy race of people, called Bankers, who subsist by fishing and whaling, occasionally by wrecking, and by raising for market a small, wiry, tough-sinewed, splay-hoofed variety of horse, called the bank pony, or marsh pony, which subsists on the coarse salt grasses of

> the wide marshes which margin the sound. These animals receive no care, save at the annual "penning" frolic, when the banks and marshes are "driven," as in a deer hunt, and the horses collected in hundreds in order to be claimed and branded, or sold. (Kerr, 1875, p. 15)

By the mid-1700s, Banker watermen and whalers from New England regularly hunted pods of right whales around Cape Lookout as they migrated north from their calving grounds. It appears that these short-range whaling operations were successful for more than 150 years.

As recently as 1875, the North Carolina state geologist described a thriving whaling industry around Cape Lookout:

> Whaling is carried on chiefly along the Shackleford Banks, between Cape Lookout and Fort Macon. The whales are taken in April and May, sometimes 5 or 6 in the course of one or two weeks. They are the common right whales, 40 to 60 feet [12–18 m] long; and a single animal frequently yields, in oil and bone, $1,200 to $1,900 [roughly $25,000–39,000 today]. On one occasion, two sperm whales were taken, one of which measured 62 feet [19 m] in length. (Kerr, 1875, p. 15)

After the American Revolution, shipping increased along the Atlantic Seaboard, and consequently, so did shipwrecks. A 95-ft-tall (30-m) light station was built at Cape Lookout in 1812, but the structure was too low and dim to warn ships off the perilous shoals that extended 10 mi/16 km into the ocean. In 1859, the structure was replaced by the present 163-ft/50-m lighthouse, which was outfitted with a Fresnel lens that directed its beam far across the water. In 1873 the lighthouse was painted with a bold pattern of black and white diamonds to enhance visibility. Even so, shipwrecks remained commonplace.

Shackleford is generally higher than Core Banks and has a more varied landscape, including high dunes at the western end. The dunes provide a windbreak that allowed the myrtle (*Myrica* spp.), cedar (*Juniperus virginiana*), and live oaks (*Quercus* spp.) of the maritime forest to take hold almost 2,000 years ago, which in turn offered storm protection for people and animals alike.

Bratton and Davison (1987) wrote that the larger trees were cut aggressively for shipbuilding and homesteads, and by 1819 the supply of live oak and cedar was noticeably diminished. But coastal charts drawn between 1850 and 1870 and published in 1888 show extensive maritime forest on Shackleford all the way to its connection with Core Banks. A chart issued in 1966, however, showed forest cover on Shackleford greatly reduced and confined to the western half. The 1906 biennial report of the state geologist described Shackleford Banks as once covered with dense forest within the memory of the island's oldest inhabitants. Engels, however, paints a different picture. He wrote in 1952 (p. 714)

> There is no evidence that extensive logging ever was practiced; the "cutting of timber" was limited to the immediate needs of the inhabitants. The latter were primarily fishermen; they lived on the soundward side of the island, not on the ocean beach. . . . It is unlikely that at that time, or for some time afterward, the number of sheep on the island was great enough to seriously decrease plant reproduction.

The oldest tree on Shackleford identified in a 1974 study was a cedar about 100 years old, but old trees on the Outer Banks are notoriously hard to date. Their centers are often rotted away, and during hot, dry periods they form false growth rings, sometimes five or six in a season, that can make the tree appear much older than it is.

Evidence of the dynamic nature of the barrier islands is everywhere. Layers of peat and broken stumps along the beach indicate the presence of a swamp forest about 200 years before, when sea level was lower and the beach much farther to the south. Stumps are also found in the salt marshes along the north border of Shackleford, where historic maps showed forests, indicating sea level rise.

Around 1840, dune fields on Shackleford Banks began to migrate and engulf the woods on the eastern end of the island. A photograph from 1917 shows a "ghost forest," the gnarled trunks of dead cedars protruding from sandy hillocks like fingers from a giant grave.

Livestock took the blame for creating migratory dunes refractory to stabilization attempts—yet by the 1970s, the island showed "extensive vegetation cover over most of the island" (Godfrey & Godfrey, 1976, p. 115) despite the continued pressure from horses, sheep, goats, and cattle. Eventually, geologists came to realize that it is a natural characteristic for some dune systems to mobilize spontaneously

A live dune on Cumberland Island National Seashore engulfs and kills vegetation in its path. Sand dunes periodically mobilize and migrate, then stabilize and revegetate, a natural process that cannot be duplicated with earth-moving equipment and plantings.

and swallow everything in their path, independent of the activities of man or beast.

Because Barden Inlet did not separate Shackleford from Core Banks until the Hurricane of 1933, horses and other animals were free to move from one island to the next. Horses typically forage within a preferred home range and do not often travel great distances, but over time there was a genetic interchange between the Core Banks and Shackleford herds as various bands interbred and expanded their ranges.

Ruffin described them unflatteringly: "These horses are all of small size, with rough and shaggy coats, and long manes. They are generally ugly. Their hoofs, in many cases, grow to unusual lengths. They are capable of great endurance of labor and hardship" (1861, p. 132). He also spoke of horses introduced to the herd "from abroad" that were not as hardy as the Banker horses, commenting "they seldom live a year on such food and under such great exposure" (1861, p. 132). He added,

> In applying the term *wild* to these horses, it is not meant that they are as much so as deer or wolves, or as the herds of horses,

A pair of Corolla Horses threads a maze of sand fences employed to slow the attrition of the primary dunes. In some places on Currituck Banks, natural dunes are in a "live" migratory phase, to the vexation of developers. Many blame free-roaming horses for destabilizing the dunes, but dune migration is an inevitable process.

wild for many generations on the great grassy plains of South America or Texas. A man may approach these, within gun-shot distance without difficulty. But he could not get much nearer, without alarming the herd, and causing them to flee for safety to the marshes, or across water, (to which they take very freely,) or to more remote distance on the sands. Twice a year, for all the horses on each united portion of the reef, (or so much as is unbroken by inlets too wide for the horses to swim across,) there is a general "horse-penning," to secure, and brand with the owner's marks, all the young colts. The first of these operations is in May, and the second in July, late enough for the previous birth of all the colts that come after the penning in May. If there was only one penning, and that one late enough for the latest births to have occurred, the earliest colts would be weaned, or otherwise could not be distinguished, as when much younger, by their being always close

At a Cape Lookout pony penning in 1946, men and boys crowd the fence to get a closer look at the restless horses corralled within. The horses appear to have Spanish characteristics, and they bear a strong resemblance to their modern relatives on Shackleford Banks and Ocracoke, as well as the horses pictured in the 1910 roundup near Oriental, N.C. (facing page) Photograph by Aycock Brown, courtesy of the Outer Banks History Center.

to their respective mothers, and so to have their ownership readily determined.

The "horse-pennings" are much attended, and are very interesting festivals for all the residents of the neighboring main-land. There are few adults, residing within a day's sailing of the horse-pen, that have not attended one or more of these exciting scenes. A strong enclosure, called the horse-pen, is made at a narrow part of the reef, and suitable in other respects for the purpose—with a connected strong fence, stretching quite across the reef. All of the many proprietors of the horses, and with many assistants, drive (in deer-hunters' phrase,) from the remote extremities of the reef, and easily bring, and then encircle, all the horses to the fence and near to the pen.

This postcard, produced around 1910, shows a roundup along the Neuse River near Oriental, N.C. Free-range horses and cattle still roamed much of Pamlico County and elsewhere on the mainland. These small horses appear to have Colonial Spanish characteristics and some coat colors typically found in the Banker herds. They are probably very similar to the horses originally released on the Outer Banks to graze. The ancestors of both groups probably included Chickasaw or Seminole Spanish horses. Note the high corral fence, suggesting that these small horses could jump a lesser barrier. Courtesy of University of North Carolina at Chapel Hill (*Poney* [sic] *Penning on the Beach*, ca. 1910).

There the drivers are reinforced by hundreds of volunteers from among the visitors and amateurs, and the circle is narrowed until all the horses are forced into the pen, where any of them may be caught and confined. Then the young colts, distinguished by being with their mothers, are marked by their owner's brand. All of the many persons who came to buy horses, and the proprietors who wish to capture and remove any for use, or subsequent sale, then make their selections. After the price is fixed, each selected animal is caught and haltered, and immediately subjected to a rider. This is not generally very difficult—or the difficulties and the consequent accidents and mishaps to the riders are only sufficient to increase the interest and fun of the scene, and the pleasure and triumph of the actors. After the captured horse

> has been thrown, and sufficiently choked by the halter, he is suffered to rise, mounted by some bold and experienced rider and breaker, and forced into a neighboring creek, with a bottom of mud, stiff and deep enough to fatigue the horse, and to render him incapable of making more use of his feet than to struggle to avoid sinking too deep into the mire. Under these circumstances, he soon yields to his rider—and rarely afterwards does one resist. But there are other subsequent and greater difficulties in the domesticating [*sic*] these animals. They have previously fed entirely on the coarse salt grasses of the marshes, and always afterwards prefer that food, if attainable. When removed to the main land, away from the salt marshes, many die before learning to eat grain, or other strange provender. Others injure, and some kill themselves, in struggling, and in vain efforts to break through the stables or enclosures in which they are subsequently confined. All the horses in use on the reef, and on many of the nearest farms on the main-land, are of these previously wild "banks' ponies." And when having access to their former food on the salt marshes, they seek and prefer it, and will eat very little of any other and better food. (Ruffin, 1861, pp. 131–132)

Shackleford is situated where hurricanes and other large storms frequently strike. Looking at a map, one will note that the coastline is oriented from southwest to northeast from northern Florida to Cape Hatteras. On reaching the Beaufort area, the shoreline runs in almost an east-west direction. This means that a broad expanse of ocean lies to the south and southwest of Shackleford and Cape Lookout. This positions the Outer Banks for direct hits from hurricanes that have strengthened over the warm Gulf Stream. These storms rotate counterclockwise, exposing Core Banks to the unmitigated onslaught of frenzied wind and waves. Situated perpendicular to Core Banks and advantageously positioned in the lee of the longer island, Shackleford is exposed to a direct strike from the south, but it is protected to a degree from storms from the east.

The year 1893 was particularly bad for storms—at least five major tropical storms or hurricanes swept Shackleford Banks, including one in August that killed 18 people and another in October that killed 22.

The wild horses of Shackleford Banks in this 1948 photograph appear very similar to their modern-day descendants and look well-nourished. Photograph by Aycock Brown, courtesy of the Outer Banks History Center.

In 1896, two major hurricanes brushed Shackleford Banks, convincing a number of residents to relocate to the mainland.

On August 17, 1899, the great San Ciriaco hurricane flattened Shackleford. Named for the saint's day on which the storm made landfall on Puerto Rico, the storm struck San Juan as a Category 4 hurricane, flogging the island with estimated sustained winds of 135–140 mph/217–225 kph. In the days before satellites and radar, most residents were unaware that the storm was coming, and 3,369 drowned or were killed by flying debris and mudslides.

After passing north of the Dominican Republic and Haiti, the mighty storm grazed Florida, then strengthened and hit North Carolina as a Category 3 hurricane. At least 25 died in North Carolina, and Buxton clocked sustained winds at greater than 100 mph/161 kph, with gusts up to 140 mph/225 kph.

On Shackleford, seven ships wrecked. The storm washed homes off their foundations, and destroyed crops and gardens. Carcasses of dead horses and sheep lodged in trees, and families watched in horror as the waves unearthed deceased family members and scattered their bones.

On Ocracoke, 100 horses and cattle drowned. The storm surge grabbed two porpoises and lodged them in an oak tree. A U.S. Weather Bureau employee assessed the damage soon after the storm:

> The chief force of the storm was spent on the Banks between Cape Lookout and Cape Hatteras, and the waves sweeping into Ocracoke Inlet nearly flooded the towns of Portsmouth and Ocracoke. Some houses were destroyed, blown over, or moved from their foundations, but no loss of life occurred in either town. Many cattle and other domestic animals were drowned undoubtedly. Many of the "Banker Ponies" roaming wild on Core Bank, opposite Carteret county, were also drowned, but not all; in fact the relief committee that visited this section observed numbers of marsh ponies all along the Banks, as well as cows, sheep, and some flocks of geese, which seem to have found places of safety. (Moore, 1899, p. 4)

The *Atlanta Constitution* reported,

> It is now thought that the total drowned will run close to 100 if it does not overreach it . . . fully sixty to seventy houses four or five churches and numerous stores, barns and warehouses were either washed away or damaged beyond repair, and as a result numbers are homeless and destitute and others have lost crops and flocks. . . . There were several thousand wild ponies on the banks which divide the ocean from the sounds and nearly all these were drowned. ("Terrible Record of Recent Storm," p. 2)

The devastation was too much to bear. After the storm, even the most tenacious residents of Shackleford Banks decided to pack up what was left of their belongings and move to stable ground. Many settled in the Promised Land section of Morehead City or on Harkers Island. Those lucky enough to have intact homes—some 50 to 70 families—floated them on barges to places such as Beaufort and Broad Creek.

By 1902, Shackleford was deserted. The island was once again left to wildlife and free-roaming livestock. At the turn of the last century, Shackleford reportedly supported a vast number of wild horses.

> On Shackelford's banks alone are the little ponies referred to. It is strange but true, that they are found in their wild state nowhere else. There are said to be about twelve hundred of them on the banks. Inquiry made of residents as to whether the number of ponies had decreased during the past fifty

On Cedar Island, as on Shackleford Banks, foals often swim from the day of birth, and use shallows as trails as much as they would use an overland route.

> years brought the response that it had, and that until about 1850 the ponies increased. . . .
>
> The colts are covered with hair several inches in length, a nature's protection against the weather. This is called colt hair, and looks like felt. It falls off in large flakes. Most of the colts are of a faded brown color, but are sometimes black. . . . Their instinct is remarkable. They know by means of it the way to get to the mainland or to islands with the minimum amount of swimming, and the writer has seen them wade great distances without getting out of their depth, making various turns and changes of direction to conform to the shoals, yet they are fearless swimmers.
>
> Though an inlet only about two miles in width separates Shackelford's banks from Bogue banks, yet the ponies never go to the latter banks, nor do they cross the Ocracoke inlet. (Olds, 1902, p. 384)

In 1933, Barden Inlet severed Shackleford from Core Banks. William Engels, a zoologist who camped on Shackleford for a month, wrote that in 1939 "there were only two buildings on the island, each a one-room shed, the one just west of The Mullet Pond, the other on Wade Shore" (1952, p. 705). When he returned in 1952, he mentioned a summer cottage built on Wade Shore, and a hunter's camp east of Whale Creek Bay. He also encountered horses:

> Horses, cattle, sheep, goats, pigs and house-cats live on the island in a semi-wild state . . . . The horses have been famed since colonial times as the "bank-ponies"; they are popularly supposed to have reached the outer banks through the wreck of an early colonial Spanish vessel. The ponies do not enter the woodland, but roam the grassland eastward of Whale Creek Bay, and frequently visit the numerous marshy islands (1952, p. 721).

Cape Lookout somehow escaped the commercialism and population growth of the Cape Hatteras area. At one time, developers considered building a bridge to Shackleford and developing the island as a tourist destination, but the state began acquiring land in the 1960s, and ultimately it became part of Cape Lookout NS. Under the management of the Park Service, the island is evolving toward a more natural state.

Congress created Cape Lookout NS in March 1966, extending from Ocracoke Inlet in the north to Beaufort Inlet in the south, but it would be another 20 years before the seashore was fully established. Mostly undeveloped and accessible only by boat, the seashore comprises four barrier islands that buffer 56 mi/90 km of the central coast of North Carolina. Unlike most other parts of the Outer Banks, Cape Lookout is poised to cope with sea-level rise in the most natural of ways. Because there are no buildings, jetties, seawalls, or other man-made structures to block sand, the islands will naturally migrate and adapt.

Cape Lookout NS is situated on the Atlantic Flyway, and is home or seasonal stopover to at least 275 bird species, including the bald eagle, peregrine falcon, and piping plover. Wildlife is everywhere. Raccoons hunt for invertebrates in the shallows, probing the tidal creeks with sensitive fingers as if reading Braille. Nonvenomous snakes, coiled like pretzels, bake on sunny logs. In the summer months, newborn loggerhead turtles emerge on moonlit nights and flipper their way to the sea while other sea turtle species forage in the adjacent waters. Feeding black skimmers unzip the glassy water with their wakes. The island also supports terns, mergansers, herons, egrets, snapping turtles, rabbits, nutria, river otters, and many other varieties of wildlife.

Three grass communities run along the long axis of the island, comprising marsh, dune, and swale. Small tangles of thick woods, branches interlaced like Velcro®, grow randomly in the hollows.

Horses graze on *Spartina alterniflora* in a large, persistent saltwater marsh on eastern Shackleford Banks.

The landscape to the west consists of tall dunes, dense maritime forest, small freshwater marshes, and, on the western tip, a large saltwater marsh.

Sue Stuska led a research team that studied Shackleford equine grazing behavior and discovered that horses primarily consume four plant species (Stuska, Pratt, Beveridge, & Yoder, 2009). In the fall, 78.0% of the horses' diet consists of sea oats, centipede grass, and saltmarsh cordgrass (*Spartina alterniflora*). In the winter, they eat more centipede grass, less sea oats, and much less saltmarsh cordgrass, and consume small quantities of several other plants. In the spring, horses prefer sea oats, saltmarsh cordgrass, and pennywort and consume considerably less centipede grass. In summer, almost 65% of equine foraging included sea oats, centipede grass, and smooth cordgrass. Horses eat saltmeadow cordgrass (*S. patens*) consistently throughout the year without seasonal fluctuations.

The average horse consumes about 2% of its body weight in vegetation each day. One would think that the selection of fodder is related to the nutrition and digestible energy content of the plants season by season, but this does not appear to be true. Horses appear to graze preferentially on plants that taste good to them. They prefer to eat short new growth rather than old, fibrous, relatively dry mature plants.

The horses of Shackleford flourish with the same vigor that kept their predecessors alive here for centuries. They demonstrate distinctive Colonial Spanish conformation, and average a diminutive 12 hands (48 in./1.22 m) high, though individuals raised on the mainland sometimes grow taller. Shacklefords have long, thick manes that unfurl contrarily in multiple directions. Bays, chestnuts, sorrels, and blacks are common. Dun, buckskin, and palomino colors have been uncommon or absent since the Park Service euthanized carriers of equine infectious anemia in 1996. Many have white markings, but none is pinto. The relentless sunshine bleaches highlights into light-colored coats, and gives blacks a rusty burnish. Their muzzles are well-furred, especially in winter, when their jaws are bearded with guard hairs. Their eyes are intelligent and unguarded, emotions sparkling within them.

Bands of feral horses vary widely in size, usually 2–20 individuals. Just as there is no "average" human family, scientists disagree on what constitutes normal horse behavior, and the horses themselves sometimes disagree with the scientists. Behavioral and social norms may differ widely from herd to herd and from band to band.

In general, a herd stallion will not permit another male to remain past puberty, but sometimes makes exceptions. The typical wild horse band consists of an alpha male and up to five or six unrelated mares with their foals. Herd composition reflects habitat, food availability, and sometimes sex ratio within the population.

Harem stallions sometimes tolerate a young subordinate stallion unrelated to the mares in the harem as long as he remembers his place and does not attempt to mate with the mares. Multiple stallions are generally present only in bands of more than 9 horses. Usually, this arrangement begins when a subordinate stallion follows at the fringes of the band and defends it from intruding stallions. The patriarch initially tries to chase the interloper away, but over time skirmishes become less frequent.

In time, the harem stallion realizes that the subordinate can lighten the workload, giving him more time to graze, mate, and rest. The lower-ranked stallion assumes the task of scent-marking the urine and feces of the mares to discourage rivals. He defends the group from competing males and shepherds wayward youngsters back to the band.

A stallion and his mare watch another stallion drive his mares through a corner of their home range. Home ranges often overlap at resources such as water sources, and these stallions worked out a compromise and saw no reason to come to blows.

The beta stallion may not mate with mature harem mares, but sometimes mounts the patriarch's adolescent daughters or young mares from other bands who visit during their pubescent heat cycles. Bands with more than one stallion are more stable, and mares are less likely to leave. Uncommonly, three or more stallions may share a band.

Although these arrangements are temporary, seldom lasting more than one breeding season, there are benefits for the subordinate stallion, too. When a stallion who has worked in this sort of "internship" moves on to join a bachelor band, he ranks higher in the in the social hierarchy.

In some cases, a pair of half-brothers will leave their natal herd and form an alliance, driving off rivals. When they acquire mares, sometimes the more dominant of the two accomplishes most or all of the matings in the band; at other times, the stallions divide up the mares in estrus—"you take that one, I'll take this one"; and sometimes it is a free-for-all in which either stallion mates with whoever is convenient whenever he wants.

Two-male harems were uncommon on Shackleford in the early 1980s, when the sex ratio was typically two females to every male. During the late 1990s, the ratio for the entire population was closer to 1:1. Perhaps this explains the increase in two-stallion bands. In 2005–2006, 19% of the Shackleford bands contained more than one adult stallion.

The band generally adheres to a set pattern of activity, moving along well-worn trails to locations within the home range that provide food, water, and relief from insects. The alpha mare will usually lead the herd in daily activities, and the stallion will bring up the rear; but he will move to the front if there is danger ahead. When a threat appears, the stallion will snort an alarm and the herd will mobilize, foals toward the center, stallion placing himself between his band and the menace.

Stallions gather the band using a driving posture—head low, ears flat, menacing look in the eye. Moving the lowered head from side to side, "snaking," implies extreme threat that, unheeded, is followed by a nasty bite. Rank has privileges. The most dominant horse is the first to drink and has first choice of the available food, followed by the beta, and then the subordinates in descending order of status.

Horses are accomplished at both flight and fight. Mares usually back-kick for defense or offense or to discipline an unruly foal, though they may strike with front hooves when rejecting the advances of a stallion. Stallions may kick, bite, lash out with a hoof, or rear and strike.

Ordinarily, wild horses do not defend a territory. Instead, they maintain a movable "sphere of intolerance" within a home range that overlaps the spheres of other bands. The stallion grazes his band of mares within a preferred range, attacking any rival males that violate his invisible boundaries. In the 1970s, Shackleford Banks had the distinction of being one of the only places in the world where biologists observed horses defending territories rather than simply guarding harems. The island is so narrow that home ranges spanned from sea to sound in a band across the island, and stallions

could easily spot intruders from a distance over the low dunes and grassland. (A single Assateague stallion, Comma, was observed displaying territorial behaviors on a similarly narrow, treeless stretch of island.)

The territorial stallions on Shackleford outcompeted weaker rivals and controlled home ranges with prime forage and easy access to water. Mares associated with dominant males grazed for an average of two extra hours per day and enjoyed increased fertility. Rubenstein was surprised to find these territorial stallions maintaining very large harems of mares. While his studies were in progress, around 1980, an increase of bachelor males upset the social order by overthrowing the older harem studs. These younger stallions divided the mares, and the territorial system dissolved.

A study involving Przewalski Horses found that the ratio between adults and juveniles in the herd strongly influenced the rate of aggression between individuals. When there are many foals in a herd, they tend to bond like gangs of naughty children and show aggressive behavior. This is probably because their mothers and other adults have less opportunity to regulate their behavior when they are off romping with playmates.

The dominance hierarchy is crucial to the social structure of the herd. When a new horse enters the band, fights ensue until the newcomer establishes a position in the hierarchy. After this initial trial, he or she generally maintains status by nonviolent threats such as pinning back the ears whenever another horse intrudes. These threats are well heeded by lower-ranking individuals, so actual kicking and biting are usually unnecessary. A more subtle manifestation of rank is the avoidance order. Lower-ranking horses will yield to dominant herd members by staying out of their way, dropping eye contact, and trying not to attract their attention. As long as a horse's position is well defined, he or she apparently feels secure, even if very low in rank.

The effects of a dominance system are to reduce overall aggression in the group and to control personal space. Status is a learned relationship between individuals, and a horse can learn to be more dominant by developing effective manipulation strategies such as bluffing.

One privilege of rank is that a dominant stallion spends more time grazing and less time defending his mares against other stallions. Dominant mares spend more time eating and have access to the highest-quality forage, have faster-growing foals, and engage in more sexual behavior. Dominant mares sometimes prevent subordinate mares in heat from mating with the stallion.

Bachelor stallions investigate a "stud pile", a pile of droppings deposited purposefully by other stallions near a watering hole. These layered manure heaps are often seen along well-traveled horse trails and can become quite large. Stallions stop and smell them thoroughly, then deposit their own manure on top and sniff again. When more than one stallion is present, the dominant horse defecates last, so that his scent may prevail.

The established hierarchy persists for as long as the group remains a stable, closed unit. Once defined, position in the social hierarchy remains relatively static until injury, old age, departure, or death of herd members changes it. An individual's place in the hierarchy is influenced by age and order of arrival in the harem, but not necessarily by size and weight. In domestic pastures, a Shetland Pony can be dominant over a Belgian draft horse.

The first mares to arrive in a harem tend to be of higher rank, and belligerent horses also rank higher. Older mares tend to rise in social standing. The sons of dominant mares tend to be higher in rank as well, but not their daughters. A new mare that joins a band is usually dominant only over the female offspring of the resident mares—unless the mares in the new herd are young and already acquainted with the new mare. Unsettled dominance relationships are mostly found between young horses and horses new to the band.

Social dynamics are usually more complicated than simple ranking. One horse may be dominant over another horse, who is dominant over a third, while the third horse is dominant over the first. As confusing as these relationships can be to human researchers, every horse knows his or her status.

Horses form cliques and friendships, keeping company throughout the day's activities. The strongest alliances are between mare and foal and within "best friend" pair bonds. Most horses have one or more friends with whom they graze, rest, engage in mutual grooming, and conduct most of their daily activities. Pair bonds can withstand periods of long separation, documented as long as 5 years in Icelandic mares. Horses of pair bonds often show jealousy when their companion mutually grooms or is courted by another horse.

When horses initiate mutual grooming, one horse will approach another with an invitation readable on its face. If the other horse agrees, each horse, beginning at the neck, gently nibbles and scratches her companion with her teeth, sometimes working her way to the root of the tail. Studies have shown that this grooming is more than pleasant scratching. It cleans the coat and tends to strengthen the social bonds between horses. Zoologists Claudia Feh and Joanne de Mazières (1993) discovered that when a horse's withers are nibbled in this fashion, her heart rate slows by 10% on average. Mares typically engage in mutual grooming with their foals, the stallion grooms with his favorite mares, foals of similar ages groom one another, and friends groom friends.

The stallion is not always the most dominant animal in the herd, especially if he is young and inexperienced. Houpt and Keiper (1982) found that all the stallions they observed were subordinate to some of the mares in their bands. Other studies including one on Shackleford showed stallions were usually the highest-ranked horse in the herd, but Keiper's study in 1986 showed that the stallions were not dominant in any of the Assateague Island bands (Keiper,1986).

The personalities of horses are as varied as those of dogs or people. Each forms friendships within the group and displays unique preferences and quirks. Like people, horses have personality conflicts. A stallion may be affectionate with one mare and bicker constantly with another. Even mothering skills differ from mare to mare. Foals from previous years, especially fillies, may remain close to their mothers.

The social dynamics of a wild horse herd are complex. Some animals form close friendships, and others are perpetually at odds. Horses within a band are usually patient and indulgent with young foals, and a mare's status rises temporarily after giving birth.

Understanding a horse's basic need for companionship, a role or status within the herd, and physical contact with other horses can make one uncomfortable to realize just how lonely and sterile are the lives of many domestic horses kept solitary in a box stall or paddock much of the time.

Just as some gentlemen prefer blondes, a stud may be attracted only to mares of a certain color, usually the color of his dam, and will even go so far as to collect a band of identically marked mares. Some stallions, wild and domestic, will refuse to mate with mares of the "wrong" coat color. Most stallions prefer high-ranking mares, show little interest in fillies under the age of 3, and avoid mating with their own daughters.

The Shackleford Banks and Carrot Island horses make their own water holes by digging in the sand with their hooves until they reach the water table. A study by Blythe (1983) compared the salt content of water from various sources on the Rachel Carson National

Beneath the sand of barrier islands, rainwater collects to form a freshwater lens. Thirsty horses need only locate where the aquifer lies closest to the surface and excavate a drinking hole, which slowly fills with freshwater when they reach the water table.

Estuarine Research Reserve and found these water holes to hold surprisingly fresh water. A tidal flat on the ocean side had 483 mEq of salt per liter of water, Taylors Creek (the narrow passage between Carrot Island and the mainland) had 477 mEq /L, and a tidal lake that connected with the sound had 339 mEq /L. But water holes dug by horses at various locations on the island had a salt content ranging from 68 mEq /L to less than 10 mEq /L. Aside from temporary rainwater pools, the holes dug by horses are the only freshwater sources on Carrot Island, and they probably provide drinking water for other wildlife.

How is it possible for freshwater to lie below the surface of an island with no permanent freshwater source? The Ghyben-Herzberg principle explains this phenomenon, which concerns the relationship between freshwater and salt water on islands and peninsulas in proximity to the ocean. The barrier islands of the East Coast and many nearby areas, such as Carrot Island, are made mostly of sand surrounded and permeated by salt or brackish water. Rainfall seeps into the sand, and a lens-shaped body of freshwater develops. This lens

The water in the holes dug by horses is muddy and fills the excavation slowly, but its salt content is minimal.

floats atop the denser salt water, "much as an iceberg floats in the ocean with most of its mass submerged" (Blythe, p. 70).

For every foot (30 cm) that this freshwater stands above sea level, it extends 40 ft/12.2 m below sea level. During periods of frequent soaking rains, the water table rises and the bottom of the lens sinks; the opposite happens during periods of drought.

To access the water table, the horses sometimes dig so deep that only their rumps are visible as they drink from the pool. Freshwater can be slow to seep into these depressions, and pony herds may spend long periods near them to ensure that each herd member gets enough. Horses return to these same holes for as long as they continue to produce, enlarging them with every visit. When water is particularly scarce, low-ranking horses may become dehydrated.

On Shackleford Banks, a large, shallow freshwater pool named Mullet Pond provides water to horses on the west end of the island. Spring rains also create seasonal freshwater pools—hoofprints along the edges advertise that the horses use this source as well.

Edmund Ruffin wrote,

> On the whole reef, there are no springs; but there are many small tide-water creeks, passing through and having their

The rainy season brings vernal pools, freshwater that collects in depressions. Horses use this water source until it evaporates or sinks down to replenish the freshwater lens that forms below the sand.

> heads in marshes, from which their sources ooze out. Their supply must be from the over-flowing sea-water. I could not learn, and do not suppose, that these waters, even at their highest sources, are ever fresh. Water that is fresh, but badly flavored, may be found anywhere (even on the sea-beach), by digging from two to six feet deep. The wild horses supply their want of fresh water by pawing away the sand deep enough to reach the fresh-water, which oozes into the excavation, and which reservoir serves for this use while it remains open. (1861, p. 133)

One 1900 account in a Scottish magazine describes Banker Horses catching and eating fish in these holes! The author explained, "they catch [fish] for themselves at low-tide, using their hoofs to dig deep holes in the sand below high-water mark; and they greedily devour the fish so left stranded, often fighting fiercely over an especially tempting one" ("Fishing Horses," 1900, p. 493). Because it is unlikely that horses would purposely catch live fish, the observer—who may have been separated by several degrees from the final publication—must have been watching horses drink from their water holes, perhaps so close to the tideline that fish were trapped within them. There is also an account of a couple of marsh ponies devouring an angler's sea bass, but this too may be a "fish story" (Clarke, 1892). On reconsideration, however, filching a fish might not be any more aberrant than raiding coolers for unlikely comestibles such as tuna sandwiches and soda pop, as do the Assateague horses.

Historically, horses grazed other islands in the area, including Hog Island, Browns Island, Harbor Island, Chain Shot Island, and Cedar Island. Steve Yeomans, a Cedar Island horseman, remembers when many of the locals would turn unneeded domestic horses out to join the herds in the marsh. Roundups were held in the summer. About 30 people would spread out to form a chain, and "walk them in." Stockmen would brand the horses, provide veterinary care, and trim overgrown hooves. They freed most back to the marshes (S. Yeomans, personal communication, October 15, 1995).

The same technique of "walking them in" was used on Shackleford Banks. An account from 1946 relates,

> Early in the morning, the beaters begin driving the ponies down the banks toward the trap, with a line of men which thinly stretches from the ocean to the sound. . . .
>
> When the ponies approach the funnel leading into the pens, the volunteers armed with leafy twigs help prevent them passing around the pen or dashing off into the sound. Once in the pen, owners apply their brands to the colts, identifying them by watching whose mares they are following. Brands are registered in Carteret County courthouse. If any bidders are present, a few may be sold, at from $50.00 to $100.00, and taken to the mainland, where they are used as pets and riders. . . .
>
> The mares and foals follow their stallion-leader unhesitatingly, and even the horse-hunters cannot restrain a cheer when a wise old banker stallion outwits [h]is adversaries. Sometimes he will climb to the top of a sea-oat-tufted dune, and sniff at the danger ahead and behind, while his herd huddles quivvering [*sic*] below, the frightened foals nuzzling close to their mares[.] At his moment of quick decision, the leader sometimes plunges headlong toward the pen, his crowd galloping blindly behind him[.] But once in awhile he will find a

These men seem mesmerized by the strength and beauty of a defiant black stallion as he tests the rope that restrains him at a 1946 penning at Cape Lookout. Photograph by Aycock Brown, courtesy of the Outer Banks History Center.

> break in the line, and wheeling swiftly, plunge into the sound, reach deep water, and swim around behind the discouraged beaters who have walked miles in the sun, only to have their quarry outwit them. (Munsell, 1946, p. 11)

Margaret Willis (1999) had similar impressions of the horses when she attended the Park Service gathers of the late 1990s. She wrote,

> These smart little horses know well how to hide just on the other side of a dune or in a little valley and within the thickets and forest. One can walk right into them without being aware of their presence until near contact or right by them then turn to see alert, soft brown eyes quietly watching.
>
> Once out in the open, these horses can disappear before your eyes then reappear some distance away, speeding in excess of 32 miles an hour, noses in the air and showing you their heels. Most seem to enjoy playing these hide and seek games, almost smiling as they slip away.

Banker Ponies had long been valued for riding and draft throughout the area. "Not all of them make good riding horses," Yeomans explained. "Some are unridable. They're just too smart. Most of them are easier to train than other breeds. Once you get them used to a saddle and bridle, they'll do anything in the world for you" (personal communication, October 15, 1995). Some were terrific swimmers and could even cross wide, swift currents in the channel separating Harkers Island from the mainland with a child astride. He told of a mare that would not flinch when a large rifle was shot from between her ears. Yeomans' grandmother lived on Shackleford Banks when livestock still roamed freely. For generations, Banker people and Banker Horses worked as partners. The horses were transportation, muscle, and recreation. Children grew up riding barefooted on bareback horses, running the beaches and swimming the sounds. For generations, for as far back as anyone could remember, the horses were there.

A 1902 article in the *New York Times* reads,

> East of Greensboro . . . it is the ambition of every child to own a Banker, and each town is likely to boast three or four of the pretty creatures. Hitched to diminutive buggies and wagons, they trot around the streets, taking steps about as long as a dog's. They are then as gentle as lambs for once tamed their tameness is absolute. ("Wild Horses of North Carolina," 1902, p. 27)

At one time, Shackleford roundups were conducted by men and boys, but during the 1950s, the pennings became part of July 4th celebrations and involved most of the community. Horses were removed to Harkers Island or the mainland for riding. Sometimes Bankers were brought over to serve as summer mounts for children. Come September, the horses were regretfully returned to the island to resume their wild lives while the children resumed their domesticated ones.

Shackleford horses are rugged and hardy, curious and smart. Carolyn Mason of the Foundation for Shackleford Horses commented, "When most horses who haven't been handled see something unusual, they often will spook and run away. A Shackleford will walk right up and investigate" (personal communication, May 25, 2009).

In 1960, an edict by the 1959 General Assembly of North Carolina required stockmen to remove all livestock from Shackleford Banks. A news account reported,

> The legislation instructed the Sheriff of Carteret County to shoot them, unless their owners herded and removed them to the mainland. The "head" count is estimated to be 500 sheep and goats and 25 head of cattle on the Island.

Owners had constructed a corral to complete evacuation, but along came violent hurricane Donna to blow it away. The Sheriff has not commenced firing. But there is little time left. The move is aimed at conservation, to permit grass to grow and anchor the sand blown inland from the Atlantic ocean [*sic*].

Still, the natives and the tourists who look for the "Banker Ponies" in the area more than goats and cows, and forward to the roundup, will miss this attraction. ("Take 'Em Alive," 1960, p. 4)

Stockmen did remove large numbers of livestock in the early 1960s, but apparently not all. The remainder multiplied, and between 1978 and 1981 the livestock census on Shackleford ranged from 81 to 108 horses, 64 to 89 cattle, 104 to 144 sheep, and 100 to 150 goats. By that time there were no year-'round residents on Shackleford, but landowners maintained fishing cabins and other structures for seasonal use.

In 1986 the Park Service razed the remaining buildings and allowed natural processes to resume. It requested that stock owners remove sheep, goats, and cattle from the islands because the grazing of feral ungulates appeared detrimental to saltmarsh and grass-shrub areas and to dune-stabilizing grasses. Then the agency slaughtered any remaining livestock "to prevent the spread of disease" (Saffron, 1987).

Local residents, however, strenuously resisted the removal of Banker Horses from Shackleford Banks. Park Service resource management specialist Michael Rikard said in an interview in the New Bern *Sun Journal*, "For some reason, there's very strong public sentiment for the horses, as opposed to goats and pigs" (Gengenbach, 1994, p. C1).

Cape Lookout NS Superintendent Preston Riddel planned to decrease the herd to a "representative number," professedly for its own good (Saffron, 1987). "The horses, for their own sakes, would be better off somewhere else," he said. "It's a very, very hard life. They belong in an environment where they can be protected." He went on to assert that if left to themselves, herds of wild horses would inevitably increase until the vegetation could no longer support them—disregarding the fact that they had been in residence for centuries.

In the 1980s, biologists determined that horses primarily consumed *Spartina alterniflora*, saltmarsh cordgrass, which readily recovered. Goats, on the other hand, browsed relentlessly in the maritime forest and had a negative impact on the ecosystem. Ultimately, in 1987, the Park Service allowed a representative herd of feral horses to remain on Shackleford "because of their potentially historic origin" (Prioli, 2007).

Lacking sheep, goats, and cattle to compete with them for resources, horses multiplied rapidly, from a relatively stable count of roughly 100 from the 1970s to 1986 to an estimated high of more than 221 in 1994. A management crisis had developed. The Park Service declared that the equine population overgrazed the island and strained the ecosystem. Rubenstein (1982) wrote, "Grazing competition on Shackleford is very intense. Despite the fact that horses spend over 75% of each hour grazing, bodily condition remains poor, and juvenile death rates remain high" (p. 484).

Mares who bear foals at the extremes of reproductive age are particularly vulnerable to malnutrition, as are foals and yearlings of either sex. Nature's way of keeping the balance is to allow the most vulnerable horses to die of starvation when food sources are exhausted. It was feared that if the Park Service did not intervene, large numbers of horses would suffer preventable deaths, and in the meantime certain native plants could also be grazed out of existence. Rikard said "We are concerned about a major die-off of horses. . . . Some out there are real thin and looking bad" (Park service [sic] study," p. 3).

The Park service had reason for concern. In the mid-1980s, a wild horse die-off struck Carrot Island, a small, marshy island lying to the west of Shackleford. Carrot, Town Marsh, Bird Shoal, Horse Island, and Middle Marshes make up the 2,315-acre/938-ha Rachel Carson NCNERR, one of 27 reserves around the country designated for research, education, and stewardship. The four reserves in North Carolina represent the diversity of habitats in the state's estuarine ecosystems. Of these, Rachel Carson and Currituck both reluctantly tolerate horses.

Carrot Island appears on the Moseley map (1733), and it was the site of a fishery in the early 1800s. In 1782, during the Revolutionary War, a small British party landed near the mouth of Taylors Creek, exchanged fire with locals, and then withdrew to Carrot Island. The

In the early to mid 1990's, many of the horses on Shackleford banks were in poor condition, especially the lactating mares.

next morning, the British overcame the local troops in Beaufort and briefly occupied the town.

A fishery existed on Carrot Island as early as 1806. In the 1920s, the U.S. Army Corps of Engineers dredged Taylors Creek and deposited the spoil on Carrot, building it higher and increasing its stability. Carrot Island, Town Marsh, Bird Shoal, and Horse Island, acquired for the reserve in 1985, total more than 3 mi/4.8 km in length and less than 1 mi/1.6 km in width. Middle Marsh, acquired in 1989, is roughly 2 mi/3.2 km long and less than 1 mi/1.6 km wide. The Reserve was named in honor of Rachel Carson, who conducted research at the site in the 1940s.

In the late 1940s, a Beaufort physician named Luther Fulcher, who also owned horses on Shackleford Banks, released six of his horses to graze Carrot Island and its associated salt marshes, intertidal flats, and tidal creeks. These animals were probably not the first equids to live there. Paula Gillikin, Rachel Carson site manager, said, "Horses were likely on the property long before the 1940s; although at this time it cannot be proved as there isn't enough documentation. . . . Horse Island has been on the maps since the 1800s. It probably got that name because at some point there were horses there" (personal communication, March 11, 2013).

Amy Muse (1941, p. 69) wrote of the early 1900s,

> At some elusive period early in this century, Beaufort changed considerably. Banker ponies were prohibited on the Town Marsh and Bird Shoal, so they were no longer able to swim across the channel at low tide to graze along the sidewalks or run up and down the streets at night. . . . Dr. Maxwell came out with his Maxwell automobile in 1911, and from then on the familiar two-wheeled carts drawn by banker ponies began to disappear from the streets.

In July 1976, about 40 acres/16 ha of the island were almost auctioned off for development. Beaufort residents who enjoyed the wild beauty of the island and its horses took action. After a legal battle, the Nature Conservancy, aided by funds raised by concerned local residents, paid $250,000 for Carrot Island and Bird Shoal.

With nothing to curb their fertility, the free-roaming horses proliferated and overgrazed the marsh. By 1986, the horse population on these small islands and marshes had reached 68. There simply was not enough food for all of them, and palatable plant species began to disappear until minimal resources were available to other species.

Unchecked, populations of large herbivores can proliferate quickly until they reach carrying capacity, the maximum number that the environment can support, whereupon the numbers typically stabilize at or below that level and the herd stops growing. A population at carrying capacity is unhealthy for both the herbivore and the vegetative community. At maximum population density, plant diversity suffers, and certain plant species may become locally rare or extinct. Likewise, an ungulate population maintained at carrying capacity has a greater incidence of disease and malnutrition, reproductive challenges, and a higher death rate compared to a herd in balance with its habitat.

A population overshoot occurs when the census temporarily exceeds the carrying capacity, resulting in overpopulation until more deaths or fewer births restore the balance. A population overshoot may precede a large-scale die-off, which occurs when the demand for food so greatly exceeds supply that the animals fail to reproduce and die in great numbers. On St. Matthew Island, Alaska, a group of 29 reindeer introduced in 1944 multiplied to 6,000 within 19 years—many more animals than the island could support. Weakened

The Rachel Carson Estuary is unable to support a large population of horses. A smaller herd, however, stays healthy, ranging across the shifting islands to access forage and water sources.

by starvation and disease, the population plummeted to 50 during one especially severe winter. All survivors were female, and the herd became extinct. Similar explosions and die-offs have occurred with reindeer on the Pribilof Islands of Alaska; sika deer in Japan and on James Island, Md.; cottontail rabbits on Fishers Island, New York; and moose on Isle Royale, Michigan, in Lake Superior.

The 68-horse herd on Carrot Island experienced a similar die-off during the winter of 1986–1987. Abundant rainfall in the late 1970s caused lush vegetative growth on Carrot, allowing the well-nourished herd to expand through increased fertility and greater survival of the oldest and youngest members of the herd. The severe drought that followed this abundance killed or stunted plants, and suddenly there was not enough food for the enlarged equine population.

Moreover, the Army Corps of Engineers dredged more than 150,000 tons/136 million kg of silt and sand from Beaufort Channel and deposited it atop the island, forming an 18-foot-high (5.5 m) dike that prevented the horses from accessing a freshwater pond. Barry Holliday, an Army Corps of Engineers official, explained that dumping sand was necessary to eliminate ground cover used by small mammals that prey on marsh birds, acknowledging that brush and grass vital to the horses was interred. "'By pumping dredge material on the island from time to time, it basically purges small bushes and shrubs where predators hide,' he said. The island's ground cover, he said, was never meant to sustain a non-indigenous species like the horses" (Saffron, 1987).

To shore up his position, Holliday reportedly added that "dumping of dredged sand created the island in the first place" (Saffron, 1987). One wonders how this dumping affected species of concern. One also wonders how horses could possibly be more damaging to the island than burying native plants and animals with dredge spoil. And it remains a mystery how dredging begun in the early 20th century, in part to reverse the natural growth of Town Marsh, could have created a recognizable Carrot Island, bearing that very name, nearly 200 years earlier (Moseley, 1733).

Famine, disease, and parasites killed 29 horses within a few months. Once concerned locals realized what was happening, they brought in hay as supplementary feed; but the starving horses, accustomed to native grasses, were reluctant to eat it. Spring brought numerous foals, and by August 1988 the herd numbered 51. The North Carolina Division of Marine Fisheries had helped them through the 1987–1988 winter by providing 20 bales of hay each week. A point well ensured adequate freshwater. No horses died that winter, which was significant; typically, during an equine population collapse, about 75% of foals die, and almost all the yearlings. But clearly this level of human assistance could not continue. If the horses were to remain on the island, managers must ensure that their numbers remain in balance with their environment.

Biologists determined that Carrot Island could comfortably sustain between 15 and 25 horses, and in 1988 the state removed 33 of 52. Nine of the 33 removed tested positive for EIA and were euthanized. Private individuals adopted the remainder. "To avoid another population collapse, protect herd health, and minimize environmental damage, the herd size is currently managed through a birth-control program similar to those administered by the National Park Service at Cape Lookout and Assateague Island National Seashores," says Gillikin (personal communication, April 18, 2011). Additionally, the Army Corps of Engineers currently deposits dredge spoils *alongside* Carrot Island rather than on top of it.

Birth control for wild horses is accomplished through immunocontraception in many eastern and western herds. In 1988, wildlife reproductive specialists Jay Kirkpatrick and John Turner, with help from Irwin Liu of the University of California at Davis, developed a contraceptive that tricks a mare's immune system into blocking sperm from receptor sites on the zona pellucida, the transparent, noncellular protein layer that surrounds all mammalian egg cells. Using the zona pellucida of pigs, they created an inexpensive, evidently harmless, and mostly reversible injection delivered by dart gun that essentially vaccinates wild mares against pregnancy without capture or restraint.

In 1998, the Secretary of the North Carolina Department of Environment and Natural Resources permitted a representative herd of 30 horses to remain on the Rachel Carson Reserve. The reserve keeps a record book that tracks each horse, noting parentage, appearance, reproductive record, contraceptive doses, general health, social habits and, eventually, death. Occasionally, horses have wandered to places where their presence is even less welcome In 1994, three bachelor stallions swam from Carrot to Radio Island—off limits for horses.

Public opinion has influenced the policies that have allowed the herds to remain, but people still voice concerns. Personnel at the reserve have received requests to plant forage for the horses to eat, offers from local businesses to buy supplemental feed, appeals to evacuate horses to the mainland before storms, and opposition to immunocontraception and Coggins testing.

Some horse advocates argue for a hands-off approach. They support leaving these horses to live entirely as wild animals with minimal interference from people, despite challenges such as limited freshwater and exposure to weather extremes including heat waves, ice storms, nor'easters, hurricanes, and floods. Others demand that the horses be managed with all the accoutrements of domesticity. But if they are to be sheltered, vaccinated, fed, wormed, and given farrier

By the mid-1990s, the Carrot Island herd had apparently recovered from its population crash, and the horses grazing across from the Beaufort waterfront appeared well-nourished.

care, the reserve might as well take it one step farther and offer them for adoption on the mainland.

A herd large enough to maintain genetic diversity without periodic introductions of unrelated horses is too large for the food resources in the Rachel Carson Reserve. The Reserve comprises several islands separated by shallows, creeks and mud flats that frustrate access by personnel and make immunocontraception difficult to implement. In the 1990s, veterinarians repeatedly attempted to test each Carrot Island horse for EIA within the same week and were unsuccessful because of the inaccessible terrain.

Horses remain on Carrot Island, but relations with their caretakers are strained. The state of North Carolina does not view wild horses as repatriated native wildlife, but as invasive exotics. The managers of the state lands on which they graze see the horses as incompatible with the management goals of the reserve and fear long-term ecological consequences. They believe that grazing and trampling of marsh grasses will accelerate erosion, and sea-level rise could further limit resources. Yet, ironically, Rachel Carson herself, namesake of the reserve wherein the horses roam, did not consider horses harmful to coastal wildlife refuges. She wrote—on behalf of the U.S. Fish and Wildlife Service—that the 300-odd cattle and ponies grazing the roughly 9,000

Horses congregate on the grassy arms of Horse Island, where the sea breeze keeps flies at bay. The author waded from Carrot Island through a quarter mile (400 m) of shallows to approach close enough for photographs.

acres/3,642 ha of Chincoteague National Wildlife Refuge were not detrimental to the waterfowl for which the refuge was established (1947, p. 17).

Reserve policies mandate maintenance of these herds, but their presence conflicts with the core mission of the reserve. Taggart, an assistant professor of environmental studies at the University of North Carolina at Wilmington wrote (2008, p. 187),

> Among the Atlantic Coast herds, conditions at the Rachel Carson site are least accommodating for the animals. With a combination of pertinent research results plus 20 years of site-specific management experience as a basis, I argue that feral horses of the Rachel Carson site should be removed for programmatic, ecological and humane reasons.

One concern is the plight of the crystal skipper (*Atrytonopsis* new species 1), a federally protected butterfly that ranges only along Bogue Banks from Fort Macon to Emerald Isle, Radio Island, and the eastern end of the Rachel Carson Reserve. This unique insect prefers untouched open-beach dune habitat and revegetated dredge spoil areas with a variety of plants. The larvae of the crystal skipper eat only seaside little bluestem (*Schizachyrium littorale),* a plant that appears unable to establish itself on the reserve, ostensibly due to pressures from grazing horses. Horses do consume small amounts of little bluestem, but mostly during the winter months, when butterflies are not reproducing.

In 2009, the reserve established four 8 ft x 8 ft/2.4 m x. 2.4 m plots and planted them with seaside little bluestem to encourage breeding butterflies. Three of the four plots were edged with pony-proof fencing, and the fourth was left open to monitor the effects of equine disturbance. As it turned out, the horses browsed the plot only once, and the plants recovered fully. No skippers used the plants, but their avoidance had nothing to do with grazing.

Nonetheless, the presence of horses on the Reserve creates a management conflict for its managers. "The Rachel Carson component of the North Carolina Coastal Reserve and National Estuarine Research Reserve is protected to provide opportunities for long-term research, education, and interpretation," says Gillikin:

> Essentially, this means that the site is managed to remain as close to a natural state as possible. Horses remain on the Rachel Carson Reserve due to the strong public sentiment attached to them. Humane management of herd size is essential to protect herd health and the natural environment. (P. Gillikin, personal communication, April 18, 2011)

At first, the North Carolina State University College of Veterinary Medicine administered a hormone-blocking vaccine to limit herd growth. This treatment was ineffective, so reserve officials began to dart mares with porcine zona pellucida vaccine in 1999. The birth-control program is the only regular intervention in their lives.

Gillikin closely monitors the equine census. In corresponding with the author, she wrote,

> As of April 18, 2011, there are 31 adult horses and 1 foal present in the herd. At this time, there are 7 harems and no bachelor bands; our bachelors are generally loners. It is not known if any mares are pregnant, as we do not perform pregnancy tests. Visual observations are used to assess pregnancy; no mares appear to be pregnant at this time. (Personal communication, April 18, 2011)

While the Rachel Carson Reserve horse population crested, crashed, and restabilized, the National Park Service took a hard look

Today the horses on the Rachel Carson Reserve appear healthy. Horses in this small, isolated population are similar in size, color, and conformation because of a shrinking gene pool.

Contraceptive vaccines limit the number of new foals and maintain the herd at a level healthy for both the environment and the horses.

at the equine population dynamics at Cape Lookout National Seashore. By the late 1990s, Park Service personnel voiced concern that the Shackleford Banks herd would suffer a die-off similar to what had occurred on Carrot Island if they did not take action to decrease the population. The agency used research by Gene Wood and Daniel Rubenstein to better understand the relationship of the horses to the native wildlife and vegetation. Data in hand, the Park Service pondered how best to handle the equine population dilemma.

Barrier island ecosystems are intricate, naturally maintaining a dynamic equilibrium of biology, geology, and physical processes that have remained in balance for tens of thousands of years. Researchers who attempt to study a single component of this ever-changing web have difficulty extracting meaningful data from the lively interplay. The scientific method can usually prove that change has taken place, but can only speculate on causation and key influences. When managers manipulate habitat, wild species often respond in an unexpected manner for reasons that are not immediately obvious. Our conclusions are invariably generalizations, because ecologists cannot yet accurately model the relationships of organisms and environment in detail. Moreover, we often filter conclusions through our expectations and biases.

Land managers around the world use managed grazing to maintain a healthy environmental balance. Godfrey and Godfrey (1976, p. 115) wrote that Shackleford's horses graze the *Spartina* in the salt mashes "down to within an inch of the mud," but concede that it is difficult to accurately assess the overall impact of grazing livestock on the Outer Banks. "Some localized areas were undoubtedly overgrazed and thus livestock were blamed for the 'deteriorated' condition of the entire Outer Banks" (Godfrey & Godfrey, 1976, p. 115).

Whether horse grazing is ultimately helpful or harmful to the environment depends largely upon herd size. Overpopulation is clearly detrimental to horses and habitat alike. By the mid-1990s many of the horses in the Shackleford herd were underweight, particularly the mares, and researchers were seeing signs of increasing environmental damage. It appeared that Shackleford Banks had more horses than it could comfortably support.

Whenever the management of wild horses is up for discussion, the enigmatic pull horses have on our emotions tends to polarize people

Whereas overgrazing can damage the environment, light grazing creates disturbances favorable to many native animals and plants.

for or against their continued residency. Heated arguments erupt between advocates and detractors, many of them armed to the teeth with hard facts, half-truths, conjecture, and strong feelings. Scientific data does little to resolve issues rooted primarily in culture, economics, and politics.

In the 1980s and 1990s, the Park Service saw free-roaming horses as uninvited guests at a dinner table set to nourish native wildlife. The agency proposed several possible management plans, including simply removing the horses from Shackleford Banks to allow the island to regain its former character. According to Park Service policy, "management—up to and including eradication" is warranted if a species deemed exotic "disrupts the accurate presentation of a cultural landscape, or damages cultural resources" (USNPS, 2006, p. 48). Paradoxically, the Park Service concedes that the horses are themselves a cultural resource. Thus the agency can neither deport them nor keep them without neglecting part of its mission.

Many Down-Easters—residents of the lowlands and villages of eastern Carteret County—considered the horses an important part of part of their cultural heritage and wanted them to remain. Many had grown up on the backs of the rugged little Bankers, and had fond memories of participating in annual pennings. Comments on the Seashore's 1982 *Draft General Management Plan* revealed strong public support for the horses, and a commitment to maintaining the herd on Shackleford Banks. The *Final General Management Plan* specified that a representative herd would remain on Shackleford Banks after federal acquisition was complete.

The Park Service considered removing a number of horses from the island and "managing" the rest. The agency proposed that a herd of roughly 60 horses could be self-sustaining if outside genes were periodically added to revitalize the gene pool. A roundup could be held and the surplus horses adopted by the public. Stallions could be castrated. Mares could be maintained on contraceptives. Another workable alternative would have been to restore part of the herd to its historic range on Core Banks and limit numbers though contraception, but the Park service vetoed this option.

A contingent of local people favored controlling the population through annual roundups. Horse pennings had been a tradition on these islands for centuries. As on Chincoteague to the north, an annual roundup could stimulate tourism, and proceeds from the sales of young stock could support horse management on the island. The Park Service opposed this plan and argued that removal of foals alters the social and reproductive dynamics of natural horse behavior. Horse penning would trample the vegetation disrupt the environment in the vicinity of the pens, and place humans and horses at risk of injury. This last concern was well-founded: wild horse gathers in the American West often result in equine injury and death. In 1998, the Park service gathered the Shackleford herd with a helicopter and hunting dogs to test for disease and euthanized two injured horses.

In 1994 officials from the Cape Lookout NS held a series of open meetings with the public and interested groups at the North Carolina Maritime Museum in Beaufort. Surprisingly, more than half of those present felt that the horses should be removed or, if permitted to remain, prevented from breeding and simply left to live out their natural lifespans. After 20 or 25 years, the horses would be gone.

Ultimately the Park Service decided to gather all the horses, test them for EIA with the assistance of the North Carolina Department of Agriculture, destroy any positive reactors, and offer the remainder for adoption. The Park service would return 50–60 horses to the island and use contraception vaccines to limit fertility. With the

A shaggy winter coat does little to hide the bones jutting from this lactating mare. This photo was taken on the east end of Shackleford in October, 1994, when wild horses, in peak condition from summer grazing, should have reserves enough to sustain them through a lean winter.

birth rate in check, the population would never again exceed what the island could comfortably sustain. Both the horses and their environment would be healthier.

EIA, also known as swamp fever, a disease that affects only equids, is caused by a lentivirus—a retrovirus with a long incubation period—similar to the ones that cause AIDS in humans and feline leukemia. The disease was identified in France in 1843, and it was probably first diagnosed in the United States in 1888. It occurs worldwide, but predominantly in warmer climates, and there is no vaccine or cure. It is not contagious, or spread directly from one animal to another; but it is infectious, transmitted by vectors. Horseflies can transfer blood from infected hosts into healthy animals. So can people when they use the same syringe to draw blood from multiple horses. After an incubation period of 2–4 weeks, infected horses may show signs such as fever, weakness, weight loss, jaundice, lack of coordination, and swelling of the legs and underbelly. Some may die.

Once infected with EIA, horses remain infected for life. Some positive testers are highly infectious, others are less so. Horses that survive the initial attack or contract a mild case become asymptomatic carriers that can infect others. The viral content of blood samples from positive-testing horses may differ by a factor of 1 million (Cordes & Issel, 1996). An inapparent carrier may carry an infinitesimal amount of virus, but under stress may have a flare with a viral load high enough to infect a whole herd with a few drops of blood. Cordes and Issel (1996) state that 1 ml/0.03 oz of blood from a horse with chronic EIA having a feverish episode contains enough virus to infect 10,000 horses.

In 1970 Dr. Leroy Coggins developed a serologic test for antibodies specific to the virus. The spread of EIA has been slowed by regulations that require testing of domestic horses before crossing borders, breeding, racing, entering a horse show, or any other formal activity that will bring horses from one location into contact with horses from another. Most adult horses that have a positive Coggins test are inapparent carriers. They show no obvious signs and have very small amounts of the virus in their blood, but they remain reservoirs of the disease.

Horseflies feed by slashing open the skin and lapping the blood. When a horse feels the bite, he attempts to dislodge the fly by twitching his fly-shaker muscle, stamping, biting, shaking his mane, or switching his tail. Meal interrupted, the fly alights again and makes another slash or zooms off to the next victim with wet blood on her mouthparts. More typically, the fly feeds on a single horse and remains nearby after its meal; the disease does not spread easily. The virus does not live long on the horsefly carrier, and the disease can be passed only between animals that are near each other. Whereas a single fly may spread the disease from an infected horse with a high viral count, transmission from an inapparent carrier may require 25 or more blood transfers. In the summer, there is no shortage of vectors: a wild horse may receive more than 1,000 horsefly bites per hour. Mosquitoes do not spread the disease.

If mares test positive, their foals will initially test positive as well. These foals may be infected or may simply carry passive antibodies to the disease obtained through the mare's colostrum. When mares are stable, inapparent carriers of EIA virus, the vast majority of their foals will eventually test negative, even when weaned at 5–8 months of age in areas with high populations of insect vectors. A three-year study of

Shackleford Banks

Choctaw and Cherokee horses in Oklahoma revealed that 97% of the foals born to infected mares eventually tested negative for the disease.

The Humane Society of the United States, U.S. Representative Walter B. Jones (NC 3), scientists and horse advocates vigorously opposed EIA testing of the Shackleford Banks herd. Local residents petitioned to block Park Service interference with the island horses. If EIA existed on Shackleford, they argued, it had been there for a very long time, and the herd had developed an ability to live with the disease. Clearly, if overpopulation was an issue, the presence of the disease was not limiting herd growth, and most of the horses appeared to be in good flesh and lived long lives. Horses in the asymptomatic stage are usually healthy overall, and most carriers are entirely asymptomatic.

Many of the Down-Easters resented the Park Service takeover of their ancestral lands. Their families had used these barrier islands for generations, and they felt they had a right to continue using them as they always had, gathering livestock and fishing from seasonal fishing shacks. They argued that the horses had been there "forever" and had the right to continue living on Shackleford. They accused the Park Service of manipulating data to be rid of the horses. Many believed that the push for EIA testing was a ploy to justify destroying the animals.

The people rose up and pushed back. Jerry Hyatt of, Newport, N.C., collected 1,700 signatures on a petition in favor of greater protection for the horses. He organized a rally at the Carteret County Courthouse in Beaufort, and hundreds of Down-Easters turned out in support of the horses. Said Hyatt in a letter to the editor of the New Bern *Sun Journal* (1996, p. A11),

> In a conversation with Assistant Superintendent Chuck Harris and Dr. Michael Rickard [*sic*] at the Cape Lookout National Seashore office on Harkers Island, I was told that, "The National Park Service does not care about the people of Carteret County, or their heritage, only the people of the United States. . . .
>
> They say they have held public meetings. Yes, they have. But at a time when working people could not attend, in a place with no parking and advertised only locally. Did the Park Service want public opinion, or did they "go through the motions" as required by the law?

A well-nourished foal sleeps on a dune on the west end of Shackleford Banks in 1995. While wild foals begin grazing in the early weeks of life, most of their calories come from milk for the first half of their first year. When resources are scarce, unweaned foals usually appear healthy, while their mothers become lean and vulnerable to illness.

Rikard maintained that the Park Service had no intention of eliminating the herd. "We're not trying to get rid of the horses. We're just trying to control the population," he said (Gernert, 1996, p. A3). Protests and pleas did not deter the Park Service. On November 12, 1996, the agency corralled all 184 horses—considerably fewer than the original estimate of 221. A Coggins test was performed on every horse. Seventy-six horses tested positive for EIA. Sixteen of the 18 dominant herd stallions on Shackleford Banks—89%—tested positive. None of these horses showed obvious signs of the disease.

Because no one can predict the risk posed by any given infected horse over time, veterinarians take the conservative position and assume that each infected horse poses the same threat at all times. North Carolina state law mandates the destruction or quarantine of any horses with positive Coggins tests to prevent them from infecting others. Because the disease can be spread only by fresh blood, and blood dries rapidly on the mouthparts of biting flies, horses are considered quarantined if they are stabled 200 yards/183 m from other horses.

The Park Service made plans to euthanize all the horses with positive Coggins tests. Down-Easters were horrified. Most positive testers showed no outward sign of illness and appeared in robust good health. The horses who were inapparent carriers could pass the virus only to herdmates. And the risk of transmission is exceedingly low: only one horsefly out of 6 million is likely to pick up and transmit EIA virus from an inapparent carrier. In other situations, the Park Service practices nonintervention in the natural processes of wildlife, even in the presence of disease. These horses were on a barrier island and posed no threat to mainland horses. Were they not *already* quarantined?

Coggins testing of feral horses was not a federal mandate. On Assateague Island NS, horse advocates pointed out, feral horses were *not* tested, despite being in fairly close proximity to domestic horses that campers bring to the island for beach riding (though not during fly season). Because many of the Chincoteague National Wildlife Refuge horses tested positive in the 1970s, it is quite possible that the untested Maryland herd harbors positive reactors. The Park Service recognizes that EIA may exist in that herd, but views it as a natural disease and does not intervene.

Down-Easters tried to find an alternative to euthanasia, protesting that the Park service disregarded their input and concerns. Carolyn Mason, a retired librarian committed to protecting the culture and heritage of the area, organized local residents to form the Foundation for Shackleford Horses, a 501(c)(3) nonprofit corporation.

The foundation proposed isolating the infected animals on Davis Ridge, a remote island-like hammock of 1,200 acres/486 ha in Core Sound and Jarrett Bay, connected to the mainland by marsh. Representatives from the Park Service, the N.C. Department of Agriculture, and the North Carolina Horse Council investigated the site, but eventually rejected it because of lack of security for the horses and inaccessibility to state veterinarians who would need to examine them periodically. No other practicable options surfaced, but officials did not allow much time.

North Carolina state law stipulates that any equid testing positive for EIA be quarantined for 60 days before retesting; a second positive result "likely means a death sentence" (Prioli, 2007, p. 62). Yet on November 20, 1996, only 8 days after the horses were corralled, the N.C. Department of Agriculture and the Park Service killed the

This magnificent black stallion, Dionysius, was gathered in the roundup of 1996 and released back to the island as a healthy horse. Other herd members were not so fortunate. In 1996, 76 Shackleford Banks horses tested positive for EIA and were subsequently euthanized.

76 positive testers "in a clandestine middle of the night debacle" in Clinton, N.C. (Willis, 1999). Also euthanized was an uninfected foal who slipped into the quarantined herd to remain with its dam. Their bodies were unceremoniously buried in a landfill.

The 108 horses with negative Coggins tests were released back to the island after the Park Service freeze-branded large numerals on their rumps to make identification easier. Freeze-branding involves applying a supercooled instrument to the horse's skin, which damages the pigment-producing cells (melanocytes) without causing the greater tissue injury of hot branding. The shape of the brand eventually grows in as white hair.

Several horses evaded captors in the 1996 roundup and were not tested. A second roundup in March 1997 captured 103 horses. Five of these were positive for EIA. The foundation was ready this time and had obtained prior approval for an isolation site. The Park Service gave these animals to the foundation, sparing them from destruction. In 1998, three of 106 horses tested positive for EIA and were quarantined. Since 1997, these inapparent carriers have been maintained at an approved 192-acre quarantine site managed by the foundation and regulated by the N.C. Dept. of Agriculture. Carrying a dormant disease has robbed them of freedom, but they found a comfortable life in quarantine, courting attention from the volunteers who bring hay and water twice daily and companionably grooming one another in the shade.

At that point it was still possible that annual roundups would continue to eliminate EIA carriers until only a few horses remained. The quarantined horses might have been the only survivors of a rare strain of an ancient breed. Since almost no offspring of positive testers carry the disease, the foundation considered the possibility of breeding the quarantined horses to preserve bloodlines should the wild herd become extinct—but the state required the castration of quarantined males. The 114-member wild herd was gathered again in 1999. Margaret Willis wrote of this roundup, "all but a few, mostly the old, are in excellent condition . . . AND they all tested negative for EIA" (Willis, 1999).

In September 2008, 11 years after their capture and quarantine, two horses developed clinical signs of EIA, one so severe that euthanasia became necessary. The surviving horse developed a second exacerbation the following month. In May 2009, a third horse in quarantine experienced a particularly severe acute episode and was euthanized.

In February 1996, shortly before undertaking the scheduled gathers for EIA testing, the Park Service arbitrarily decided that the Shackleford Banks component would maintain a herd of 50–60 horses through immunocontraception. This census goal had no science to support it, and it represented a compromise only between allowing the horses to set their own maximum population and

The Foundation for Shackleford Horses saved the lives of five horses that tested positive for EIA by finding them a home where they could live in quarantine.

removing them entirely. In February 1997, Rep. Jones, introduced H.R. 765, the Shackleford Banks Wild Horses Protection Act, and on August 13, 1998, President Clinton signed An Act To Ensure Maintenance of a Herd of Wild Horses in Cape Lookout National Seashore (16 U.S.C. §459g–4, 2010), which ordered the Interior Department to "allow a herd of 100 free roaming horses" and forbade removal unless the population exceeded 110.

It also mandated a partnership between the Park Service and the Foundation for Shackleford Horses, Inc. (or another qualified nonprofit entity), for the management of free-roaming horses in the seashore. Equine geneticist Dr. E. Gus Cothran evaluated the markers present in the DNA of the horses and concluded that the herd had excellent genetic variability—but for how long?

In the early days of co-management, the relationship between the Park Service and the foundation was often turbulent. Rubenstein wrote,

> Debate often turned rancorous over disagreements on how best to manage this historical horse population. In particular, too much effort was expended in trying to interpret at what

In the 1990s, the Park Service freeze-branded large numerals on the left rump of the Shackleford horses that were re-released to the island.

> threshold management should begin and end, as well as on balancing the mechanisms of selective removal and fertility control. (*H.R. 1521, H.R. 1658 and H.R. 2055*, 2003a)

In October of 2002, a new superintendent, Bob Vogel, took the reins of Cape Lookout NS and organized a meeting that included scientists, representatives from the foundation, and other stakeholders to rethink the existing management plan. The team reached a consensus that the horse population should never fall below 110 horses and that occasional expansions to 130 animals would allow successful genes to increase in frequency and benefit the population. In May 2003, Rep. Jones introduced H.R. 2055 to give this determination the force of law. Although legislation never reached the full Senate in the 108th Congress, it was a defining moment because an agency of the Department of the Interior publicly endorsed increasing the size of a wild horse herd. P. Daniel Smith of the Park Service spoke to the House Committee on Resources in support of increasing the number of horses in the herd. He said, "The Department is strongly committed to conserving, protecting, and maintaining a representative number of horses on the Shackleford Banks portion of the Seashore, as we have done in other units of the National Park System which contain horses, and believes that the number of horses on Shackleford Banks

Sporting comically fuzzy ears and an undisciplined Shackleford mane, a young filly named Sprite dozes in the shade near her dam. Immunocontraception allows herd managers to restrict herd growth with minimal disruption to the animals. Very few foals are born on the island each year, most of them to mares with uncommon bloodlines.

should be determined by the ecology of the island and by means which protect the genetic viability of the Shackleford Banks horses" (*H.R. 1521, H.R. 1658 and H.R. 2055*, 2003b).

Jones tried again in January 2005 with H.R. 126, sometimes referred to as the Cape Lookout National Seashore Free-Roaming Horse Law Amendment. In October 2005, Pete Domenici (NM), chairman of the Senate Committee on Energy and Natural Resources, reported, "The range of 110 to 130 horses is based on sound science and provides the population changes, which are necessary for maintaining the genetic viability of the herd" (S. Rep. No. 109-154, 2005). The recommendation became Public Law 109-117 in 2005 (An Act . . . To Allow for an Adjustment in the Number of Free Roaming Horses Permitted in Cape Lookout National Seashore, 2005).

The author was surprised to learn that through this legislative process, the Department of the Interior had unequivocally supported, and Congress had eventually approved, a proposal virtually identical to one for improving the genetic health of the dangerously inbred

Corolla herd, which both entities have repeatedly rejected. Says Karen McCalpin of the Corolla Wild Horse Fund,

> H.R. 306 (formerly H.R. 5482), the Corolla Wild Horse Protection Act, mirrors the Shackleford Banks Act with one important exception—it allows for the introduction of mares from Shackleford Banks. This would immediately breathe new genes into our dying gene pool. . . .
>
> The Shackleford horses live on 3,000 acres, have been managed at a target of 120–130 (with never less than 110) for the last 12 years, and with no documented negative impact to the National Park. The Corolla horses have access to nearly 8,000 acres. Only a third of that is owned by the Department of Interior—the rest is private land. It is not an issue of lack of carrying capacity to support 120–130 horses. (McCalpin, 2011)

Cedar Island, across the sound from Core Banks, is perhaps best known as the southern terminus of a state ferry to Ocracoke. It is also home to a small, little-known herd of wild horses. Although Cedar Island NWR (a low-profile operation with no resident staff) takes up more than half the island, the horses live on private land. A series of low islands and marshes owned by longtime residents has functioned for years as a wild horse sanctuary. The range lies east of the ferry dock and extends for about 8 mi/13 km, in a swath about 2 mi/3.2 km wide. Sponenberg (2011) writes that the original Cedar Island herd comprised horses taken from Core Banks, "which were supplemented by a later addition of Ocracoke horses." For more than a century, perhaps much longer, herds of 100–200 bays, chestnuts, buckskins, and blacks lived wild in these marshes, sharing the abundant forage with other wildlife and with cagey feral cattle that charged when surprised. After helping the Park Service euthanize nearly half the Shackleford herd in unseemly haste, the Veterinary Division of the N.C. Department of Agriculture turned its attention to the few wild horses left on Cedar Island and nearly eradicated them as well.

"Local people used to round up the horses every 4th of July," says Nena Hancock, who manages the Cedar Island herd with her husband, Woody. "Many people came to help and to watch. Some of the horses were branded by their owners during the roundups and

While the Department of the Interior unequivocally supported maintaining the Shackleford herd at a biologically sustainable 110-130 horses, the same agency has inexplicably balked at setting the same parameters for the genetically compromised Corolla herd, even though the Currituck horses range over a much larger area. As a result, the Corolla horses are increasingly inbred and are in danger of extinction.

others were sold" (personal communication, March 9, 2011). Locals attest that horses have roamed Cedar Island for more than a century, their numbers augmented in 1958 by a group of horses removed from Portsmouth Island. An Associated Press article documented the translocation:

> Those shaggy wild things called Banker ponies which roam this off-shore North Carolina island, will get their annual penning-up next Friday.
>
> The island animals are a herd of the famed outer banks ponies. In previous years some of them have been sold to private owners after the annual roundup, but few of them will be sold this year. The Cedar Island Banker Pony Assn. which looks after them, said plans are to build up the herd.
>
> The pony herd was moved to Cedar Island, 12 miles east of Atlantic from Portsmouth Island several months ago. The transfer was ordered by the 1957 General Assembly in an effort to halt beach erosion. ("Round-Up Slated for Banker Ponies," 1958, p. 10B)

White Sands Stable, owned by Wayland Cato, offered boarding and trail riding along the beach near the Cedar Island ferry dock. Equestrians from all over brought their horses to White Sands for the opportunity to ride on the beach. In 1996, one of these visiting horses apparently brought EIA to the stable, where it spread rapidly to infect all 11 of Cato's domestic horses. Seven of these horses were quarantined in Virginia, and the rest were destroyed.

In June 1997 veterinarians from the N.C. Department of Agriculture gathered the wild herd from the Cedar Island marshes and tested each horse for EIA. Of the 15 wild horses, 13 tested positive for EIA and were euthanized. "When the original herd became sick, Woody and Clyde helped the state vets round up the horses and test them," says Hancock. "After the testing was done there were only two remaining mares."

The wild horses of Cedar Island were a beloved part of local history and culture. Island residents were determined to save the herd and introduced genetically similar horses from Shackleford Banks to replace those that had been euthanized. "Woody and Clyde asked the local land owners, who agreed to allow us to reintroduce the horses," says Hancock (personal communication, March 9, 2011). "The land is still owned by several different people who graciously allow us to keep the wild herd of horses around, as they have been there for as long as the local people remember. Because we help with the Shackleford horses, we were able to get a stallion to reintroduce with the two remaining mares. Over the next couple of years, the Shackleford foundation gave us several more mares to restart the herd."

New foals are named for elderly or deceased Cedar island residents—Becky, Kassie, Ronald, Ulva, Ina May. Bucky, the lone buckskin mare, is the sole survivor of the original herd. She has produced a number of foals, and in 2010, she delivered Gay, a lovely buckskin filly very much like herself.

By 2010, there were 39 horses in the Cedar Island marshes, 6 of them gelded, most between the ages of 8 and 12. Thirty to 40 wild cattle also roam these marshes, as they have for more than a century. Steve Edwards writes "They are managed by people that care greatly about them. This is no BLM "round'em up and lock 'em up story. The briefest conversation with Woody Hancock is all that it takes to see

Bucky, a buckskin mare whose mane and tail bleach to blonde in the sun, was the last survivor of the original Cedar Island herd. When horses were brought in from Shackleford to resurrect the herd, Bucky found a mate, and 2010 in she produced a filly named Gay. The original bloodlines will continue.

where his heart is. He wants only the best for these horses" (2011, February 23).

Despite strong local support, argument continues about whether the grazing of wild horses on Cedar Island and Shackleford Banks damages the environment. At the northeast end of Cedar Island, two actively migrating barrier spits with grassy dunes extending into southern Pamlico Sound mirror the migration mechanism of barrier islands. The northwestern and southeastern spits share similar tidal dynamics and sand supply, but have been divided by a sturdy fence for 45 years. The free-roaming horses and cattle of the island use the southeastern spit. The grazed spit has shorter growth of *Ammophila* and *Spartina* grasses, low or absent dunes, and broad overwash fans, while the ungrazed area is lush, with tall, grassy dunes and narrow, localized overwash channels. Could the light to moderate grazing occurring now cause changes such as these, or were they a result of historic overgrazing when horse and cattle populations were higher? More research is necessary to answer this question.

Cedar Island wild horses graze under a leaden sky. The horse at the far left is "Shack," otherwise known as Deliops (*spoiled* in reverse), the 1998 son of the elderly mare #16 of Shackleford, pictured elsewhere in this chapter. He has sired numerous foals in the privately owned horse sanctuary on Cedar Island.

Private citizens have managed the Cedar Island herds humanely, cooperatively, and at no cost to the public for many years. In contrast, federal law required Cape Lookout NS and the Foundation for Shackleford Horses to manage the Shackleford herd jointly beginning in 1998, but real cooperation took years to develop.

The Park Service now makes an apparently sincere effort to integrate public opinion and information into its management plans. From a distance, Sue Stuska delivers contraceptives by dart gun, documents new foals, and evaluates the health status of each individual, but does not feed, touch, or interact with the horses except in extraordinary circumstances. Cape Lookout NS treats them as wildlife and grants them the space in which they can be wild horses. It does not provide veterinary care, farrier care, or vaccines. If a horse is suffering from a terminal condition, they decide whether to euthanize on a case-by-case basis.

The PZP contraceptive vaccine has been in use on Shackleford since January 2000. In January, Stuska's team performs pregnancy tests on each mare's manure. They administer PZP vaccines between

Horses are social animals, but sometimes build community in surprising ways. One Cedar Island mare does not typically interact with other horses, but instead prefers the company and protection of three large bulls that share the marshlands. In heat she consorted with a stallion, but after conceiving her foal, she retreated to the security of her bovine guardians.

late February and April, beginning when a mare is 2 years old. The vaccine reduces pregnancy by 97% the first year and 76% following an annual booster in the second year of treatment. The side effects of immunocontraception include an occasional small abscess at the injection site, a higher rate of late season births, and changes in the genetics or social order of the herd. The advantages are better health and longer lives for contracepted mares, and increased foal survival in a herd kept in balance with its environment. By vaccinating individual mares, the Seashore saves the entire herd from the stress of repeated gathers.

Horses are selected for removal or contraception based on matrilineage (how many horses represent the mare's line) and genetics. The pedigrees comprise at most 4 generations and trace back to any of 47 horses identified at the onset of record keeping for whom the Park knows neither parent. In reality, all the horses on the island are probably related to one another in some way.

An article in the Cape Lookout NS publication *Preserve and Protect* asks, "If you were managing a barrier island population for the

Mare number 16, age 22 at the time of this 2010 photograph, shows loss of muscle on the neck and prominent spine and hips. These age-related changes are normal, exacerbated by the nutritional stress of nursing her most recent foal, 15U, and the eight offspring that preceded her. All in all, she appeared in good health. She died in 2012 at the age of 24, a good long life for a multiparous wild mare. Her son Deliops was moved to Cedar Island, where his healthy genes breathed new life into the dwindling Cedar Island herd.

future, and wanted to be sure it had the best chance to adapt to changing conditions, what attributes would you choose?" ("Managing wildlife for a changing ecosystem," 2008, p. 7). On Shackleford, herd managers have made preserving the herd's diverse gene pool a priority so that when conditions change, some animals within the population should have whatever adaptive genes will be needed.

The seashore staff bases its decisions about contraception and removal on the number of representatives of a line, individual factors such as whether a horse is socially and physically able to reproduce, and a concept called "mean kinship" ("Managing wildlife," 2008, p. 7). Mean kinship is a number assigned to a horse that shows how closely it is related to other members of the herd. A foal born to well-represented parents would have many aunts, uncles, cousins, or siblings and would be assigned a higher mean kinship number than one born from rare lineages. A foal with a high mean kinship number is a likely target for contraception or for removal from the island, especially if it has a number of full siblings.

Managers limit the reproduction of these horses so that the fertile family will not eclipse the rarer lineages over time. As of 2008, seven lines were represented by a single horse, five by two horses, and nine by three horses ("Managing wildlife," 2008, p. 7). "We don't have a policy to let every mare reproduce," says Stuska. "We look at each horse as an individual" (personal communication, May 26, 2010). She pointed to a wall chart depicting the lineage of each horse. One family line had more than 30 representatives.

Number 79, a sorrel stallion, is the only representative of his line. His branch of the family tree will probably end with him because he has been unable to acquire and keep a mate and is unlikely to do so. Because records have been kept for only a short time, however, and because paternity can be uncertain, he may well have unidentified aunts, uncles, or cousins on the island, and they may perpetuate some of the same genes.

Assateague Island NS uses a different approach, allowing the horses themselves to decide which lines prevail and which extinguish (J. Kirkpatrick, personal communication, May 29, 2014). On Assateague, each mare is given the opportunity to reproduce and make a genetic contribution to the herd. Opportunity does not always equal success, and the drama unfolds in often surprising ways. For example, the offspring of the unusually prolific mare M4 failed to reproduce, and her once well-represented line ended. On the other hand, the N4 and N6 lines became dominant despite limiting each mare to a single foal. Kirkpatrick explains, "Certain lines will be more successful than others—that's natures's way and the stuff of evolution. By trying to manipulate genetics through contraception you may very well be shorting the most successful lines in the herd and trying to inflate the least successful" (personal communication, May 29, 2014). Eggert et al. (2010) studied kinship and pedigree using DNA in fecal samples of Assateague horses and found that although mitochondrial DNA diversity (inherited through the mother only) is low, nuclear DNA remains as diverse as that in established breeds. After considering various strategies to maximize genetic health, they concluded that the current method of allowing each mare to reproduce once optimally preserves the long-term viability of the herd.

The wild horse management team keeps track of each individual in the herd and its habits, home ranges, and preferred companions. This black mare of the "D" matrilineage, famed for her distinctive low-set ears, lives on the far west end of the island.

Research has repeatedly shown that immunocontraception does not alter daily activity patterns, social relationships, or harem fidelity. Other studies seem to contradict some of these findings and conclude that contracepted mares are more likely to switch harems, undermining the cohesion among females. "When a mare has a foal," says Stuska, "her whole focus is to eat, eat, eat to meet the tremendous energy requirements of pregnancy and lactation. She tends to stay within a harem because she has no energy to spare straying to another" (personal communication, May 26, 2010). A vaccinated mare, on the other hand, spends a considerable amount of time soliciting and receiving sexual interest. "Perhaps she instinctively realizes that if this stallion isn't impregnating her, she should follow her biological mandate to find another mate who will," she added (personal communication, May 26, 2010). A stallion is unlikely to impregnate a contracepted mare, so in theory she may move from band to band, unsatisfied.

Earlier research demonstrated that the age and experience of the harem stallion is the primary influence on band stability. Assessments by Gray (2009) and Turner (2011), both cited in NRC (2013), showed no change in band fidelity among PZP-vaccinated mares

Horses form bonds that can last for a lifetime. This Cedar Island pair is inseparable. They lived together on Shackleford Banks as mare and stallion, and she bore him a foal. They were moved to Cedar Island to replace horses lost to an epidemic of EIA, and in time herd managers gelded the stallion. Even with reproduction no longer a possibility, the bond endures.

in Nevada and on Assateague Island, respectively. It is difficult to compare the results of studies involving different designs, methods, objectives, environments, time frames, and variables. Disagreement among researchers and confusion among readers is therefore understandable. After looking at available research, the National Research Council (2013, p. 110) concluded,

> The importance of harem stability to mare well-being is not clear, but considering the relatively large number of free-ranging mares that have been treated with liquid PZP in a variety of ecological settings, the likelihood of serious adverse effects seems low.

Some social behavior patterns hold true for most wild horses; others are unique to individuals or their herds. Although some researchers have reported that wild equine harems show little change in composition from year to year, others have found that band fidelity differs widely from population to population. Before immunocontraception was initiated in the Carrot Island herd, about 30% of the mares changed bands in late winter. During a 5-year study on Cumberland Island, about 38% of the mares remained in their original harems, and the majority of mares changed bands from one to four time).

On the one hand, mares who live in unstable harems weigh less, carry more parasites, and have less time for grooming or grazing, and their foals are more likely to die. On the other hand, immunocontraception decreases foal mortality in the herd as a whole and extends the life of mares, which probably balances any negative effects of changing harems.

Maintaining a smaller herd has benefitted the horses greatly. Rubenstein (1982) wrote that Shackleford foals had only a 48% chance of surviving the first two years. Since 2000, about 82% of Shackleford foals have survived the first 2 years (S. Stuska, personal communication, March 8, 2013). Stuska et al. (2009) indicate that the diet of the Shackleford horses is adequate to maintain relatively good health.

Some Shackleford horses are ribby, with angular hips and narrow necks. Most of the horses that appear excessively thin are elderly, and, like many people of advanced years, often have trouble maintaining weight because of poor dentition and digestive issues. Body fat reserves are used as an energy stash in case the animal needs to burn it for fuel in times of physical stress. If a horse does not have these reserves and energy is needed, the animal will break down muscles and burn protein for fuel.

The Henneke Body Condition Scoring System is one scale scientists use to assess body condition in horses. This tool yields an accurate assessment of condition regardless of breed, body type, sex or age. Assessors press on the horse's body to obtain accurate information—or in a wild setting, use visual inspection only. Each body area is assessed and scored, and the numbers are totaled and divided by 6.

Any given horse's ideal score varies with breeding role and workload. For endurance and polo horses, the ideal Henneke score is 4–5, whereas dressage horses and show jumpers should score 5–7. A breeding stallion should score 4–6, and a pregnant mare should score 6–8 (Kentucky Horse Council, n.d.). A mare with a score of less than 4.5 at foaling is considered thin and is less likely to successfully produce a foal the following year. Pregnant and recently delivered mares often become thinner over the topline because the weight of the foal pulls the flesh tighter over the back and ribs. The presence

Athletic wild stallions can grow ribby during the breeding season. Whereas a malnourished horse would show muscle wasting, this stallion has a well-developed neck and chest and is lean and fit for successful combat.

of fat deposits and well developed muscle along the neck, shoulder, withers, and tail head more accurately indicate good condition, while muscle wasting indicates the horse has run out of carbohydrate and fat resources and is breaking down muscular protein. Aged horses will have lower scores as their muscle structure softens.

Dr. Alberto Scorolli used body condition scoring to assess the general nutrition, health, and growth rate of the wild equine population in Tornquist Park, Argentina (Scorolli, 2012). His team rated the condition of each horse using an alternative method, the visual body condition score, with a scale from 0=very thin to 5=obese. He found that the condition of the horses showed an annual cycle, peaking in late summer and early fall. He found that stallions had higher values than adult females—average score 3—in every assessment—a finding consistent with studies involving other populations. Tornquist Park mares, including yearlings and 2-year-olds, had condition scores significantly lower than those of stallions; 45% of females had a BCS equal to or less than 1.5 at some point during the study, with scores highest in March and May (late summer/early autumn in the Southern Hemisphere) and lowest in September (late winter–early spring).

Henneke Body Condition Scoring System

1—Poor (Extremely emaciated; no fatty tissue)	Neck	Bone structure easily noticeable
	Withers	Bone structure easily noticeable
	Shoulder	Bone structure easily noticeable
	Ribs	Ribs protruding prominently
	Loin	Spinous processes projecting prominently
	Tailhead	Tailhead, pinbones, and hook bones projecting prominently
2—Very Thin	Neck	Bone structure faintly discernible
	Withers	Bone structure faintly discernible
	Shoulder	Bone structure faintly discernible
	Ribs	Ribs prominent
	Loin	Slight fat covering over base of spinous processes. Transverse processes of lumbar vertebrae feel rounded. Spinous processes prominent
	Tailhead	Tailhead prominent
3—Thin	Neck	Neck accentuated
	Withers	Withers accentuated
	Shoulder	Shoulder accentuated
	Ribs	Slight fat over ribs. Ribs easily discernible
	Loin	Fat buildup halfway on spinous processes, but easily discernible. Transverse processes cannot be felt
	Tailhead	Tailhead prominent but individual vertebrae cannot be visually identified. Hook bones appear rounded, but are still easily discernible. Pin bones not distinguishable
4—Moderately Thin	Neck	Neck not obviously thin
	Withers	Withers not obviously thin
	Shoulder	Shoulder not obviously thin
	Ribs	Faint outline of ribs discernible
	Loin	Negative crease (peaked appearance) along back
	Tailhead	Prominence depends on conformation. Fat can be felt. Hook bones not discernible

Score each element separately, add the scores, divide by 6, then round to the nearest ¼ point to allow for uneven fat deposit.

5—Moderate (Ideal weight)	Neck	Neck blends smoothly into body
	Withers	Withers rounded over spinous processes
	Shoulder	Shoulder blends smoothly into body
	Ribs	Ribs cannot be visually distinguished, but can be easily felt
	Loin	Back is level
	Tailhead	Fat around tailhead beginning to feel soft
6—Moderately Fleshy	Neck	Fat beginning to be deposited
	Withers	Fat beginning to be deposited
	Shoulder	Fat beginning to be deposited
	Ribs	Fat over ribs feels spongy
	Loin	May have a slight positive crease (a groove) down back
	Tailhead	Fat around tailhead feels soft
7—Fleshy	Neck	Fat deposited along neck
	Withers	Fat deposited along withers
	Shoulder	Fat deposited behind shoulder
	Ribs	Individual ribs can be felt with pressure, but noticeable fat filling between ribs
	Loin	May have a positive crease down the back
	Tailhead	Fat around tailhead is soft
8—Fat (Fat deposited along inner buttocks)	Neck	Noticeable thickening of neck
	Withers	Area along withers filled with fat
	Shoulder	Area behind shoulder filled in flush with body
	Ribs	Difficult to feel ribs
	Loin	Positive crease down the back
	Tailhead	Fat around tailhead very soft
9—Extremely Fat (Fat along inner buttocks may rub together. Flank filled in flush)	Neck	Bulging fat
	Withers	Bulging fat
	Shoulder	Bulging fat
	Ribs	Patchy fat appearing over ribs
	Loin	Obvious crease down the back
	Tailhead	Bulging fat around tailhead

Adapted from USBLM, Billings Field Office (2009, pp. 132–133).

Scorolli submits that the lower values of adult females probably reflect the caloric drain of pregnancy and lactation (Scorolli, 2012), while the weight of growing juvenile horses is influenced by both nutritional reserves and forage availability.

Body condition scoring becomes a measure of whether an individual meets its energy requirements for maintenance, growth, and reproduction, and lactation. In 1977, Eberhardt observed that as population density increases (1) the age at first reproduction increases, (2) fertility of females in their reproductive prime decreases, and (3) mortality of juveniles and vulnerable adults increases, especially through the winter (Eberhardt, 1977, 2002). Although Eberhardt's original hypothesis applied to marine mammals, subsequent research has demonstrated that it generally holds true for ungulates.

De Roos, Galic, and Heesterbeek (2009) found that nutrition and growth *in utero* and in youth are pivotal to population dynamics in free-roaming horses. Foals of well-nourished mares are born with nutritional reserves equal to that of their dams, and for the first months of life profit from additional caloric intake through suckling. If forage is abundant, they remain in good flesh through their first two winters. Conversely, a foal born to and suckled by a malnourished mare starts life at a nutritional disadvantage. When challenged to meet the demands of maintenance and growth with inadequate winter forage, these foals exhaust their bodily reserves and begin their second summers depleted. These undernourished yearlings often die over their second winter. Those that survive have a later onset of puberty and first reproduction than optimally nourished horses and face higher mortality throughout life, especially during pregnancy and lactation.

Horses grow larger and heavier when climatic conditions favor growth of forage and when the population is diffuse enough to reduce competition for resources. When a population of free-roaming horses reaches the carrying capacity of its range, the condition of both habitat and herbivores decline.

Although environmental damage is not always quantifiable in the absence of long-term studies, the body condition of horses is easy to determine, even from a distance, and it appears to give a rough benchmark of environmental conditions. Scorolli proposes using

The author took this photograph on the east end of Shackleford Banks in October 1995. There was little forage available—the tall grasses around the horses appear to be species unpalatable to horses. Many mares in this band had a Henneke score of 1.5–2. Note this mare's prominent spine, hips, and ribs beneath her winter coat, and the wasting of her neck and hip musculature. Many of these horses tested positive for EIA the year after this photograph was taken. What combination of factors led to the emaciated state of the lactating mare? Chronic disease? Repeated reproduction? Advanced age? Inadequate forage? Heavy parasite load? Regardless of cause, the lactating mares as a group had the lowest Henneke scores of all horses observed. Although thin mares produce less of milk than well-nourished mares, this mare's foal appears in reasonably good condition.

the average body condition scores of adult females in a herd as a tool to determine a herd's "health, its potential growth rate and the proximity of the population size to carrying capacity" (2012, p. 92). A more sensitive indicator, however, would be an assessment of the body condition scores of *only* the herd members under the greatest caloric stress—lactating mares.

When forage is abundant, pregnant mares typically store body fat for the first 270 days of gestation. As gestation advances, a wild mare is challenged to consume sufficient calories, and body condition will usually decline slightly. This decline continues during the first 120

The herd on Cumberland Island is unmanaged, and the equine census had apparently stabilized at the carrying capacity of the island many years prior to this 2011 photograph. The population could no longer grow because mortality balanced birth rate, and the horse count was roughly the same from year to year. When years of drought reduced the available forage, horses began to starve. Many of the lactating mares on Cumberland Island had Henneke scores of 1.5–2.5, though the author noted some individuals scoring as high as 3. Muscle wasting was evident on the gaunt mares. Where horses grazed in lawn areas, grasses were continuously cropped short with bare patches, while marsh areas showed varying degrees of impact.

days of lactation, because milk production can nearly double a mare's daily energy requirements. On average, mares produce about 3 gallons of milk daily—approximately 450 gallons/1,700 L in the first 5 months of her foal's life. Foals nurse about five times an hour for the first 5 weeks and will usually double in weight their first month of life. Burdened by this relentless need for calories, lactating mares may fail to meet energy demands when resources are scarce.

As Scorolli and others have observed, lactating mares are usually among the first herd members to lose condition when food resources are scarce. A biologist might consider a Henneke score of 4 "good" for a lactating wild mare younger than 4 or older than 15 years of age at the end of winter. By early spring, before the burgeoning of new vegetative growth, a BCS of 3 for these same mares—who could

On the western end of Shackleford Banks in May 1994, many lactating mares had Henneke scores of 2.5-3.5 but some scored as high as 4. The grass was very short throughout most of the areas favored by the horses, indicating that they were consuming the forage as fast as it could grow, yet many fell short nutritionally.

This 9-year-old Shackleford mare, 16K, photographed in May of 2009 while nursing her 4-month-old filly, scored a 4 on the Henneke scale. While her spine and ribs are discernible, her hip, neck, and shoulder muscles are well developed.

very well be pregnant again—"might not be alarming" (Kane, 2011, p. 444). If a lactating mare has a BCS of 3 at the end of autumn, her poor condition indicates that either forage has been inadequate during the growing season, or she has some other health problem. A mare that heads into winter with a BCS of 3 or less has virtually no fat reserves, and when faced with climatic challenges or famine, may break down muscle to meet energy demands.

Kane proposed that if fewer than 5% of mares are in poor condition, it is likely that those individuals have health issues, such as advanced age, dental abnormalities, chronic illness, or parasite overload. An imbalance between grazers and forage may exist if 20–30% of the lactating mares have poor BCS scores. When greater than half the nursing mares are extremely thin, the rangeland is probably inadequate to support the herd, and the potential for disastrous consequences is great. Because conception occurs in the previous growing season, some underweight mares will continue to foal until they are elderly, or their Henneke scores fall below 2. Overall, however, underweight mares have lower conception rates and decreased fetal and foal survival as compared to mares at a healthy weight.

While researching this book, the author made informal visual surveys of the condition of barrier island herds and observed a relationship between Henneke body condition scores of nursing mares and the health of herds as a whole and their environment. She proposes using the Lactational Condition Index described in the accompanying table to assess the health of herds and their proximity to the carrying capacity of their ranges. An LCI score of 1 indicates that the majority of lactating mares have Henneke scores of 1–2. Their poor condition suggests that the herd is probably at or near carrying capacity, reproductively compromised, and at risk of high mortality due to malnutrition, parasites, and disease. Herds in which the majority of lactating mares have Henneke scores of 3–4 receive an LCI score of two. These herds are generally healthier, more fecund, and have lower mortality than those with an LCI score of 1. Their ranges are at lower risk of depletion than at LCI-1, though some areas may be overused. Herds in which most lactating mares have Henneke scores greater than 5 receive LCI scores of 3 and 4. These well-nourished herds are generally healthier and optimally fecund, and their ranges are at low

On Cedar Island in May of 2010, forage remained lush despite grazing by free-roaming horses and cattle. Favored grazing areas regrew rapidly when horses moved on. All horses had Henneke scores of 5–7, including lactating mares (center) and older multiparous mares such as the grandmother of the foal (right).

risk of grazing stress. Overweight mares are the most fertile, but may develop obesity-related health problems.

Although federal legislation supports a target population of 110–130 horses on Shackleford Banks as long as they remain in balance with their environment, that number may change as the dynamic island shrinks, grows, or experiences changes in vegetation. Stuska keeps close watch on herd dynamics, utilizing a GPS system to follow the movements of bands within the herd. Key goals of management are to minimize human interference with the herd and to promote the perpetuation of genes to keep the herd optimally healthy.

In 2005, 18 horses were removed to the mainland; two joined the Cedar Island herd, and the rest were adopted. If more than 12 horses required removal, the Park Service would consider conducting a large-scale roundup of most of the herd, but no such roundups are planned.

Stuska is authorized to remove any horse from the island if necessary, such as if a horse sustains a severe injury or its life is endangered by the actions of people; if it shows "consistent, repetitious, unprovoked aggressive behavior" toward people; or if a foal is

On Currituck Banks, lactating mares typically have Henneke scores of 5–6, and the herd appears in balance with its environment. This photograph was taken in May 2012. Research indicates that the effects of grazing are temporary, and grasses recover from grazing by early summer.

On the north end of Assateague, this mare scored a 5.5 while nursing a foal in October 2009. In April, the onset of foaling season, lactating mares tend to score lower, about 3.5–4. Most mares on the Maryland end receive contraceptive vaccines annually and do not experience the stress of serial pregnancies and lactation. Though forage is abundant, favored spots sometimes show evidence of overgrazing and are slow to regrow to optimal density.

Lactational Condition Index

LCI 1

Majority of lactating mares with Henneke scores of 1–2.

Herd less likely to survive short-term insults (for example disease, drought, harsh climatic events).

Lower mare fertility and poorer pregnancy outcomes. Pregnant and lactating mares, juveniles, and elderly horses risk higher mortality.

Range at high risk for overgrazing.

LCI 2

Majority of lactating mares with Henneke scores of 3–4.

Herd more likely to survive short-term stresses.

Higher fertility, better pregnancy outcomes, greater survival.

Range at overall lower risk for overgrazing, but may show signs of overgrazing in places favored by horses.

LCI 3

Majority of lactating mares with Henneke scores of 5–6.

Herd most likely to survive short-term stresses.

Best pregnancy outcomes, high fertility, best survival.

Range at low risk for overgrazing.

LCI 4

Majority of lactating mares with Henneke scores greater than 7.

Herd most likely to survive short-term stresses.

Best pregnancy outcomes (Kentucky Equine Research, 2003), highest fertility, best survival. Horses with very high Henneke scores are at risk for obesity-related health problems.

Range at lowest risk for overgrazing.

If lactating mares, the horses under the greatest nutritional stress, remain in good condition, one can reliably expect to find the rest of the herd and the herd's environment in good condition. A low Lactational Condition Index at the end of the growing season is a telling indicator of individual and herd health. Lactational Index tool © Bonnie Gruenberg 2013, may be used freely with attribution.

Poco Latte, the smallest horse on Chincoteague NWR, scores a 5 for condition despite the fact that she gestates and nurses a foal almost every year. On the Chincoteague Refuge, lactating mares have Henneke scores of 4–5, although elderly mares can score as low as 3 or 3.5 after a severe winter. This photograph was taken in a corral during the April roundup in 2010.

orphaned before its first birthday (Cape Lookout NS & Foundation for Shackleford Horses, 2010, p 2).

If the population "blooms" beyond the target 130-horse maximum, the 2010 management plan provides for the removal of about 2–4 foals per year, mostly males. These animals are taken young, so that they might adapt easily to life in domestication, and upon leaving the island become the property of the Foundation for Shackleford Horses. These removals are a low-key affair; Stuska and her assistants typically approach the periphery of the band, quietly sedate the animal, load it into a boat, and transport it to the mainland. The band usually continues grazing, unconcerned.

The Seashore conducts an equine census each December, and that year's foals are numbered in order of birth. Stuska closely monitors births, deaths, and band composition. When the author visited in 2010, there were 114 horses on Shackleford Banks, divided into roughly 25

With a dirty look at the author, a wild Pryor Mountain mare escorts her newborn son away from human intrusion. This photograph was taken in May 2011, early in the growing season. After a tempestuous Montana winter, this mare scored a 3.5, like most of the lactating mares observed by the author on that trip.

The wild horse manager must walk a narrow line between respecting nature and maximizing the health of the herd and its environment. Severely overgrazed islands can lead to equine starvation and environmental disruption.

harems and about 7 bachelor bands (S. Stuska, personal communication, May 26, 2010). The oldest living horse was 28, and the second-oldest was 27. There were five births in 2009, although one foal died soon after birth, and the herd saw the deaths of one 8-year-old horse, two 17-year-olds, one 21-year-old and two 24-year-olds. Five foals were born in 2010. Stuska suspended contraception and removal to allow the population to expand. As of March 30, 2012, there were 109 horses on Shackleford Banks because of an unexpectedly low birthrate and higher mortality in 2011—one juvenile, five horses in their teens and two in their 20s. Six foals were born in 2011, "Y" year, and 11 in 2012, "Z" year. Two of these "Z" foals were born out of season, one in January and one December.

In 2009, four youngsters were removed from the island to be adopted. To preserve rare genes, Stuska targets a young horse for removal only if its bloodlines are well represented on the island. "If there are aunts, uncles and siblings out there, the horse is more likely to be removed. We consider whether the dam has had other offspring, and whether the grand dam has had other offspring," she explained. "Nobody knows the pedigrees of these horses for more than four or five generations back" (personal communication, May 26, 2010).

In 2012, Stuska recorded the first known horse in the herd to reach the age of 30, a mare that died the following winter. In 2013, another mare reached the 30-year milestone—#68, mother of nine. Her last colt, black stallion 4W, was born to a stallion of uncommon lineage, after her birth control vaccine failed to prevent pregnancy. She was 26 at the time. Since the Park Service started using contraception, mares typically outlive stallions on the island.

Stuska sponsors four horse-watching field trips annually for the public. She explained,

> It is important for the public to get information directly from park rangers who are knowledgeable within the subject area. On these trips, we take a boat to Shackleford Banks, walk around, look for horses, identify which ones we see, talk about what they are doing, and talk about how they are related to each other. It's a whole day of horses, walking through the marsh, braving the bugs. They always have waiting lists but you can get in if you reserve months ahead of time. (Personal communication, May 26, 2010)

Adagio, a diminutive Shackleford gelding removed from the island as a foal, was adopted in 2012. Pictured here as a yearling, Adagio loves people and followed visitors around Carolyn Mason's farm, soliciting attention like an oversized dog.

Shackleford Banks is accessible to visitors by boat; passenger ferry service is available from Beaufort and Harkers Island. The Park Service requires that visitors remain at least 50 ft/15 m from the horses—about the length of a large bus.

The Shackleford horses are relatively tolerant of people, but like all wild horses bite, kick, and charge unpredictably. The author took all the photographs in this book with a telephoto lens from a distance greater than a bus length away, but came to realize that she was still too close when she was charged by an angry stallion who suddenly spun and came after her. (Fortunately, he did not pursue when she ran away.) The prospect of injury is especially daunting when one considers the inaccessibility of the island, the significant distance from a hospital, and patchy cell-phone reception.

Banker Horses are astonishingly rugged, durable, and unflappable. Their endurance is legendary. Steve Edwards describes an informal race between Holland, his 14-hand (56-in./1.42-m) Shackleford, and a Spanish Mustang stallion descended from the legendary Choctaw Sundance. Carrying a 160-lb/73-kg rider, Holland was allowed to choose his own speed and gait. He ran 5 mi/8 km in 20:54, finishing a half mile (0.8 km) ahead of the well-bred stallion (Edwards, 2009, July 16). On a 50-mi/80-km long-distance ride, Edwards (2009, September 14) described Holland as "absolutely impeccable. He carried 220 lbs forty miles with about 35–38 of those miles at a trot and averaged 5.5 mph. I have no idea how many more miles he could have done." In a blog post Edwards wrote of Holland:

> When I ask him to go, he goes. Where I ask him to go, he goes. When we get to the briers, he goes. When we reach deep water, he goes. As far as I ask him to go, he goes. As smoothly as I asked him to go, he goes. With the ground frozen rock hard, he goes. With the ground parched and baked rock hard, he goes. With the sun glaring down on us, he goes and in the pitch darkness of the night, he goes (2011, July 17).

Shackleford horses are sometimes available for adoption. The Foundation charges a $600 adoption fee and seeks homes where the horses will be well-treated. They are highly intelligent and quickly learn to enjoy the company of people. On Carolyn Mason's farm, a Shackleford named Adagio, removed from the island as a foal, followed visitors around the yard, nuzzling and expecting to be hugged

Grinning with utter delight, the author sets out for a trail ride on Steve Edwards's Shackleford gelding Holland. Holland floated down the trail with easy, tireless gaits reminiscent of those of an Icelandic. He was remarkably surefooted, trotting and cantering over exposed roots and rocks without missing a step. He was also bomb-proof, unconcerned when deer exploded from the underbrush.

and scratched. When the inspectors for the Horse of the Americas registry came to evaluate the herd, they commented, "The domesticated ones were 'pocket ponies' that wanted to please, and were quite willing to follow one around just to get more attention" (Ives, 2007, p. 10). Shackleford Banks is loved by countless people who hold memories of the island close to their hearts. Shackleford, the Thoroughbred racehorse who won the 2011 Preakness Stakes was named for the island—his owners evidently visit Shackleford Banks frequently and find great peace and pleasure there.

The congressionally mandated partnership between the Park Service and the Foundation for Shackleford Horses has been in place since 1999. The agencies work well together, combining resources and reaching goals more effectively than either could alone. In the face of climate change and other future challenges, this management team should act to preserve both the island ecology and the herd, keeping the horses of Shackleford in a healthy balance with the island that has been their home for hundreds of years.

References

An Act To Amend Public Law 89-366 To Allow for an Adjustment in the Number of Free Roaming Horses Permitted in Cape Lookout National Seashore, Pub. L. No. 109-117 (2005).

An Act To Ensure Maintenance of a Herd of Wild Horses in Cape Lookout National Seashore, 16 U.S.C. § 459g-4 (2010).

Assateague Island Alliance. (2010). *eBay foal-naming auction.* Retrieved from http://www.assateagueislandalliance.org/name.html

Barber, & Pilkey, O.H. (2001). *Influence of grazing on barrier island vegetation and geomorphology, coastal North Carolina.* Paper No. 68-0 given at the Geological Society of America Annual Meeting, November 6, 2001. Retrieved from https://gsa.confex.com/gsa/2001AM/finalprogram/abstract_28327.htm

Bardenhagen, E., Rogers, G., & Borrelli, M. (2011). *Cape Lookout National Seashore storm recovery plan 2011: Final draft.* Washington, DC: U.S. National Park Service.

Barlow, C. (2000). *The ghosts of evolution: Nonsensical fruit, missing partners, and other ecological anachronisms.* New York, NY: Basic Books.

Barnes, J. (2007, November-December). Scattered by the wind: The lost settlement of Diamond City. *Weatherwise, 60*(6), 36–41. doi: 10.3200/WEWI.60.6.36-41

"Beach Pounders" [U.S. Coast Guard Beach Patrol personnel at Hilton Head Training Center, SC](Photograph). (*circa* 1943).

Bennett, D., & Hoffman, R.S. (1999, December). *Equus caballus. Mammalian Species, 628,* 1–14.

Bishop, E.C. (1989). *Prints in the sand: The U.S. Coast Guard Beach Patrol during World War II.* Missoula, MT: Pictorial Histories.

Blythe, W.B. (1983). The Banker ponies of North Carolina and the Ghyben-Herzberg principle. *Transactions of the American Clinical and Climatological Association, 94*(6): 63–72.

Borkfelt, S. (2011). What's in a name?—Consequences of naming non-human animals. *Animals 2011, 1*(1), 116–125; doi: 10.3390/ani1010116

Bourjade, M., de Boyer des Roches, A., & Hausberger, M. (2009). Adult-young ratio, a major factor regulating social behaviour of young: A horse study. *PLoS ONE, 4*(3), e4888. doi: 10.1371/journal.pone.0004888

Boyd, L., & Keiper, R. (2005). Behavioural ecology of feral horses. In D.S. Mills & S.M. McDonnell (Eds.), *The domestic horse: The origins, development and management of its behaviour* (pp. 55–82). Cambridge, United Kingdom: Cambridge University Press.

Bratton, S.P., & Davison, K. (1987). Disturbance and succession in Buxton Woods, Cape Hatteras, North Carolina. *Castanea, 52*(3), 166–179.

Cameron, E.Z, Setsaas, T.H., & Linklater, W.L. (2009). Social bonds between unrelated females increase reproductive success in feral horses. *Proceedings of the National Academy of Sciences of the United States of America, 106*(33), 13850–13853. doi: 10.1073/pnas.0900639106

Cape Lookout National Seashore & Foundation for Shackleford Horses. (2010). Management plan for the Shackleford Banks horse herd. Harkers Island, NC: Cape Lookout National Seashore.

Capomaccio, S., Willand, Z.A., Cook, S.J., Issel, C.J., Santos, E.M., Reis, J.K.P, & Cook, R.F. (2012). Detection, molecular characterization and phylogenetic analysis of full-length equine infectious anemia (EIAV) gag genes isolated from Shackleford Banks wild horses. *Veterinary Microbiology, 157*(3–4), 320–332. doi: 10.1016/j.vetmic.2012.01.015

Carson, R. (1947). *Chincoteague: A National Wildlife Refuge* (Conservation in Action 1). Washington, DC: U.S. Fish and Wildlife Service. Retrieved from http://digitalcommons.unl.edu/usfwspubs/1

Cheevers, W.P., & McGuire, T.C. (1985). Equine infectious anemia virus: Immunopathogenesis and persistence. *Reviews of Infectious Diseases, 7*(1), 83–88.

Clarke, S.C. (1892). Sea-bass and other fishes. In G.O. Shields (Ed.), *American game fishes: Their habits, habitat, and peculiarities; how, when and where to angle for them* (pp. 287–343). New York, NY: Rand, McNally.

Code for America Brigade. (n.d.). *Work on Adopt-a-Hydrant.* Retrieved from http://brigade.codeforamerica.org/applications/8

Conant, E.K., Juras, R., & Cothran, E.G. (2012). A microsatellite analysis of five colonial Spanish horse populations of the southeastern United States. *Animal Genetics, 43*(1), 53–62. doi: 10.1111/j.1365-2052.2011.02210.x

Cordes, T., & Issel, C. (1996, June). *Equine infectious anemia: A status*

report on its control, 1996 (APHIS 91-55-032). Washington, DC: U.S. Department of Agriculture, Animal and Plant Health Inspection Service.

De Bry, T. (1590) (Engraver). *Americae pars, nunc Virginia dicta . . .* [Part of America, now called Virginia . . .]. In T. Harriot, *A briefe and true report of the new found land of Virginia.* Frankfurt-am-Main, Germany: Theodor de Bry. Retrieved from http://memory.loc.gov/gmd/gmd388/g3880/g3880/ct000777.jp2

De Roos, A.M., Galic, N., & Heesterbeek, H. (2009). How resource competition shapes individual life history for nonplastic growth: Ungulates in seasonal food environments. *Ecology, 90*(4), 945-960.

Diefenbach, D.R., & Christensen, S.A. (2009, August). *Movement and habitat use of sika and white-tailed deer on Assateague Island National Seashore, Maryland* (Technical Report NPS/NER/NRTR—2009/140). Philadelphia, PA: National Park Service, Northeast Region.

Durham, R.S. (2009, July 6). *The mounted beach patrol.* Retrieved from http://www.army.mil/article/23935/The_mounted_beach_patrol/

Donlan, C.J., Berger, J., Bock, C.E., Bock, J.H., Burney, D.A., Estes, J.A., . . . Greene, H.W. (2006). Pleistocene rewilding: An optimistic agenda for twenty-first century conservation. *American Naturalist, 168*(5), 660–681. doi: 10.1086/508027

Dwyer, J. (2007). A non-companion species manifesto: Humans, wild animals, and "the pain of anthropomorphism." *South Atlantic Review, 72*(3), 73–89.

Eberhardt, L.L. (1977). Optimal policies for the conservation of large mammals, with special reference to marine ecosystems. *Environmental Conservation, 4,* 205–212.

Eberhardt, L.L. (2002). A paradigm for population analysis of long-lived vertebrates. *Ecology, 83*(10), 2841–2854.

Edwards, S. (2009, July 16). I could not have hit him with a shotgun. *Mill Swamp Indian Horse Views.* Retrieved from http://msindianhorses.blogspot.com/2009/07/i-could-not-have-hit-him-with-shotgun.html

Edwards, S. (2009, September 14). Lab results are in. *Mill Swamp Indian Horse Views.* Retrieved from http://msindianhorses.blogspot.com/2009/09/lab-results-are-in.html

Edwards, S. (2011, February 23). The Shacklefords of Cedar Island. *Mill Swamp Indian Horse Views.* Retrieved from http://msindianhorses.blogspot.com/2011/02/shacklefords-of-cedar-island.html

Edwards, S. (2011, July 17). Yesterday I rode a horse that was not beautiful. *Mill Swamp Indian Horse Views.* Retrieved from http://msindianhorses.blogspot.com/search?q=holland

Eggert, L., Powell, D., Ballou, J., Malo, A., Turner, A., Kumer, J., . . . Maldonado, J.E. (2010). Pedigrees and the study of the wild horse population of Assateague Island National Seashore. *Journal of Wildlife Management, 74*(5), 963–973. doi: 10.2193/2009-231

Engels, W.L. (1952). Vertebrate fauna of North Carolina coastal islands II. Shackleford Banks. *American Midland Naturalist, 47*(3), 702–742.

Evans, P. (2005, April). *Body condition scoring: A management tool for evaluating all horses* (AG/Equine/2005-01). Logan, UT: Utah State University Cooperative Extension. Retrieved from http://extension.usu.edu/files/publications/publication/AG_Equine_2005-01.pdf

Fear, J., et al. (2008, August). *A comprehensive site profile for the North Carolina National Estuarine Research Reserve.* Retrieved

from http://www.nerrs.noaa.gov/Doc/PDF/ Reserve/NOC_SiteProfile.pdf

Feh, C, & de Mazières, J. (1993). Grooming at a preferred site reduces heart rate in horses. *Animal Behaviour, 46*(6): 1191–1194. doi: 10.1006/anbe.1993.1309

Fishing horses. (1900). *Chambers's Journal, 6th Series* (3), 493.

Foundation for Shackleford Horses. (2005). *Spirit's page.* Retrieved from http://www.shacklefordhorses.org/stories/spirit.htm

Gengenbach, L. (1994, March 27). Reined in? Human population starting to threaten banks horses. *Sun Journal* (New Bern, NC), pp. C1, C5.

Gernert, T. (1996, March 15). People to protest pony control. *Sun Journal* (New Bern, NC), p. A 3.

Godfrey, P.J., & Godfrey, M.M. (1976). *Barrier island ecology of Cape Lookout National Seashore and vicinity, North Carolina* (National Park Service Scientific Monograph 9). Washington, DC: Government Printing Office.

Goodloe, R.B., Warren, R.J., Cothran, E.G., Bratton, S.P., & Trembicki, K.A. (1991). Genetic variation and its management applications in eastern U.S. feral horses. *Journal of Wildlife Management, 55*(3), 412–421.

Goodloe, R.B., Warren, R.J., Osborn, D.A., & Hall, C. (2000). Population characteristics of feral horses on Cumberland Island, Georgia and their management implications. *Journal of Wildlife Management, 64*(1), 114–121. doi: 10.2307/3802980

Goodwin, D. (2002). Horse behaviour: Evolution, domestication and feralisation. In N. Waran (Ed.), *The Welfare of Horses* (pp. 1–18). Dordrecht, Netherlands: Kluwer Academic Publishers.

H.R. 1521, H.R. 1658 and H.R. 2055: Legislative hearing before the Subcommittee on National Parks, Recreation, and Public Lands of the Committee on Resources, U.S. House of Representatives, 108th Cong. 6 (2003a) (letter of Daniel I. Rubenstein).

H.R. 1521, H.R. 1658 and H.R. 2055: Legislative hearing before the Subcommittee on National Parks, Recreation, and Public Lands of the Committee on Resources, U.S. House of Representatives, 108th Cong. 16 (2003b) (statement of P. Daniel Smith).

H.R. 2055, 108th Cong. (2003).

H.R. Rep. No. 105-179 (1997).

Harper, F. (2010, September 15). *Broodmare management in fall.* Retrieved from http://www.extension.org/pages/29125/broodmare- management-in-fall

Houpt, K.A., & Keiper, R. (1982). The position of the stallion in the equine dominance hierarchy of feral and domestic ponies. *Journal of Animal Science, 54*(5), 945–950.

Hyatt, J. (1996, March 17). After being around 400 years, Shackleford Banks herd deserves better treatment [Letter to the editor]. *Sun Journal* (New Bern, NC), p. A 11.

Ives, V. (2007). *Corolla and Shackleford Horse of the Americas inspections—February 23–25, 2007.* Retrieved from http://www.corollawildhorses.com/Images/HOA Report/hoa-report.pdf

Jones, T.H. (2004). *Cape Lookout Life Saving Station historic structure report.* Atlanta, GA: U.S. National Park Service, Southeast Regional Office, Historical Architecture, Cultural Resources Division.

Kane, A.J. (2011). The welfare of wild horses in the western USA. In W. McIlwraith &. B.E. Rollin (Eds.), *Equine Welfare* (pp. 442–462). Chichester, United Kingdom: Wiley-Blackwell.

Kathrens, G. (Writer & Cinematographer). (2001). Cloud: Wild Stallion of the Rockies [Television series episode]. In THIRTEEN (Producer), *Nature.* New York, NY: WNET.

Kathrens, G. (Director, Writer, & Cinematographer). (2003). Cloud's Legacy: The Wild Stallion Returns [Television series episode]. In

THIRTEEN (Producer), *Nature*. New York, NY: WNET.
Kathrens, G. (Writer & Cinematographer). (2009). Cloud: Challenge of the Stallions [Television series episode]. In THIRTEEN (Producer), *Nature*. New York, NY: WNET.
Keiper, R. (1985). *The Assateague ponies.* Atglen, PA: Schiffer Publishing.
Keiper, R. (1986). Social structure. *Veterinary Clinics of North America: Equine Practice, 2*(3), 465–484.
Kentucky Equine Research, Inc. (2003). *Optimal body condition scores for breeding mares* (Equine Review N21). Retrieved from http://www.ker.com/library/EquineReview/ 2003/Nutrition/N21.pdf
Kentucky Horse Council. (n.d.). *The Henneke system of body condition scoring.* Retrieved from http://www.kentuckyhorse.org/henneke-body-condition-scoring
Kerr, W. (1875). *Report of the geological survey of North Carolina.* Raleigh, NC: Josiah Turner.
King, M. (2007). *The wild horses of Shackleford Banks: An interview with Dr. Sue Stuska.* Retrieved from http://network.bestfriends.org/3448/news.aspx

Kirkpatrick, J., & Fazio, P. (2010, January). *Wild horses as native North American wildlife.* Retrieved from http://awionline.org/content/wild-horses-native-north-american-wildlife

Kirkpatrick, J. F. and A. Turner (2007). Immunocontraception and increased longevity in equids. Zoo Biology 26:237-244.

Klein, D. (1963). The introduction, increase, and crash of reindeer on St. Matthew Island. *Journal of Wildlife Management, 32*(2), 350–367.

Levin, P., Ellis, J., Petrik, R., & Hay, M. (2002). Indirect effects of feral horses on estuarine communities. *Conservation Biology, 16*(5), 1364–1371. doi: 10.1046/j.1523-1739.2002.01167.x

Linklater, W.L., Cameron, E.Z., Minot, E.O., Stafford, K.J., (1999). Stallion harassment and the mating system of horses. Anim. Behav. 58, 295–306.

Linklater, W.L., Stafford, K.J., Minot, E.O., & Cameron, E.Z. (2002). Researching feral horse ecology and behaviour: Turning political debate into opportunity. *Wildlife Society Bulletin, 30*(2), 644–650.

Madosky, J.M., Rubenstein, D.I., Howard, J.J., & Stuska, S. (2010). The effects of immunocontraception on harem fidelity in a feral horse (Equus caballus) population. *Applied Animal Behaviour Science, 128*(1), 50–56. doi: 10.1016/j.applanim.2010.09.013

Managing wildlife for a changing ecosystem. (2008). *Preserve & Protect,* 2008–2009, 7. Retrieved from http://www.nps.gov/calo/naturescience/loader.cfm?csModule=security/getfile&PageID=317930

Mayor's Press Office. (2012, January 19). *Mayor Menino invites residents to "Adopt-A- Hydrant" this winter.* Retrieved from http://www.cityofboston.gov/news/default.aspx?id=5444

McCalpin, K. (2011, January 31). Why is the federal legislation critical? *Wild and Free Weekly.* Retrieved from http://corollawildhorses.blogspot.com/2011_01_01_archive.html

McDonnell, S.M. (2005). Sexual behaviour. In D.S. Mills & S.M. McDonnell (Eds.), *The domestic horse: The origins, development and management of its behaviour* (pp. 110–125). Cambridge, United Kingdom: Cambridge University Press.

McGreevy, P. (2004). *Equine behavior: A guide for veterinarians and equine scientists.* London, United Kingdom: W.B. Saunders.

Milstein, T. (2011). Nature identification: The power of pointing and

naming. *Environmental Communication, 5*(1), 3–24.

Moore, W. (1899). *Climate and crop service of the National Weather Bureau.* Raleigh, NC: U.S. Department of Agriculture.

Moseley, E. (Cartographer). (1733). *A new and correct map of the province of North Carolina.* Retrieved from http://digital.lib.ecu.edu/1028

Munsell, J.W. (1946, August 22). Hunting and fishing. *Newark Advocate* (Newark, OH), p. 11.

Muse, A. (1941). *The story of the Methodists in the port of Beaufort.* New Bern, NC: Owen G. Dunn.

National Research Council. (2013). *Using Science to Improve the BLM Wild Horse and Burro Program: A way forward.* Washington, DC: The National Academies Press.

Noble, D.L. (1992, March). *The Beach Patrol and Corsair Fleet: The U.S. Coast Guard in World War II.* Washington, DC: Coast Guard Historian's Office.

Nuñez, C.M.V., Adelman, J.S., Mason, C., & Rubenstein, D.I. (2009). Immunocontraception decreases group fidelity in a feral horse population during the non-breeding season. *Applied Animal Behaviour Science, 117*(1), 74–83. doi: 10.1016/j.applanim.2008.12.001

Olds, F.A. (1902). The wild horse of the Banks. *Forest and Stream, 59*(20), 384. Retrieved from https://play.google.com/books/reader?id=9kMhAQAAMAAJ& printsec=frontcover&output=reader&hl=en_US

Park service [*sic*] study says wild horse herd should be thinned. (1995, September 2). *Sun Journal* (New Bern, NC), p. A3.

Pippin, J. (2005, January 11). Several wild horses to be removed from Shackleford Banks, N.C., to thin herd. *Knight Ridder/Tribune Business News.* Retrieved from http://www. accessmylibrary.com/coms2/summary_0286-7781584_ITM

Prioli, C. (2007). *The wild horses of Shackleford Banks.* Winston-Salem, NC: John F. Blair.

Ransom, J.I., & Cade, B.S. (2009). *Quantifying equid behavior—A research ethogram for free-roaming feral horses* (Techniques and Methods 2-A9). Reston, VA: U.S. Geological Survey.

Round-up slated for Banker ponies on Cedar Island. (1958, July 1). *Evening Telegram* (Rocky Mount, NC), p. 10B.

Rubenstein, D.I. (1981). Behavioural ecology of island feral horses. *Equine Veterinary Journal, 13*(1) 27–34.

Rubenstein, D.I. (1982). Reproductive value and behavioral strategies: Coming of age in monkeys and horses. In P.P.G. Bateson & P.H. Klopfer (Eds.), *Perspectives in Ethology: Vol. 5. Ontogeny* (pp. 469–487). Princeton University Press.

Rubenstein, D.R., Rubenstein, D.I., Sherman, P.W., & Gavin, T.A. (2006). Pleistocene Park: Does re-wilding North America represent sound conservation for the 21st century? *Biological Conservation, 132*(2), 232–238. doi: 10.1016/j.biocon.2006.04.003

Ruffin, E. (1861). *Agricultural, geological, and descriptive sketches of lower North Carolina, and the similar adjacent lands.* Raleigh, NC: Institution for the Deaf & Dumb & the Blind.

Rutberg, A.T. (1990). Intergroup transfer in Assateague pony mares. *Animal Behaviour, 40*(5), 945–952. doi: 10.1016/S0003-3472(05)80996-0.

S. Rep. No. 109-154, at 6 (2005).

Saffron, I. (1987, April 20). *As ponies die, an entire town feels the pain.* Retrieved from http://articles.philly.com/1987-04-20/news/26195580_1_wild-horses-dead-horses-carrot-island

Scorolli, A.L. (2012). Feral horse body condition: A useful tool for population management? In *International Wild Equid Conference, Vienna 2012: Book of abstracts* (p. 92). Vienna, Austria: Research Institute of Wildlife Ecology, University of Veterinary Medicine. Retrieved from http://www.vetmeduni.ac.at/fileadmin/v/fiwi/Konferenzen/Wild_Equid_Conference/IWEC_book_of_abstracts_final.pdf

Senter, J. (2003). Live dunes and ghost forests: Stability and change in the history of North Carolina's maritime forests. *North Carolina Historical Review, 80*(3), 334–361.

Shackleford Banks Wild Horses Protection Act, H.R. 765, 105th Cong. (1997).

Shackleford Banks horse herd update. (2013, January 7). Retrieved from http://static-horsejournal.s3.amazonaws.com/wp-content/uploads/2013/02/Shackleford-Banks-Horse-Herd-Update-2013-01-07-final.pdf

Shackleford Banks horses 2011 findings report. (2012, April 5). Retrieved from http://www.nps.gov/calo/parknews/2012-04-05.htm

Smith, P.D. (2003, June 24). *Statement of P. Daniel Smith . . . before the Subcommittee on National Parks, Recreation, and Public Lands, of the House Committee on Resources, concerning H. R. 2055, to amend Public Law 89-366 to allow for an adjustment in the number of free roaming horses permitted in Cape Lookout National Seashore.* Retrieved from http://www.nps.gov/legal/testimony/108th/capelook.htm

Sponenberg, D.P. (2011). *North American Colonial Spanish Horse update, July 2011.* Retrieved from http://www.centerforamericasfirsthorse.org/north-american-colonial-spanish-horse.html

Stevens, E.F. (1990). Instability of harems of feral horses in relation to season and presence of subordinate stallions. *Behaviour, 112*(3-4), 149–161. doi: 10.1163/156853990X00167

Stick, D. (1958). *The Outer Banks of North Carolina, 1584–1958.* Chapel Hill: University of North Carolina Press.

Stuska, S., Pratt, S.E., Beveridge, H.L., & Yoder, M. (2009). *Nutrient composition and selection preferences of forages by feral horses: The horses of Shackleford Banks, North Carolina.* Retrieved from http://www.nps.gov/calo/parkmgmt/upload/Nutrient Composition and Selection Preferences.pdf

Taggart, J.B. (2008). Management of feral horses at the North Carolina Estuarine Research Reserve. *Natural Areas Journal, 28*(2), 187–195. doi: 10.3375/0885-8608(2008)28 [187:MOFHAT]2.0.CO;2

Take 'em alive. (1960, November 16). *Burlington Daily Times-News* (Burlington, NC), p. 4.

Terrible record of recent storm. (1899, August 25). *Atlanta Constitution* (Atlanta, GA), p. 2.

U.S. Bureau of Labor Statistics. (n.d.). *CPI inflation calculator.* Retrieved from http://www.bls.gov/data/inflation_calculator.htm

U.S. Bureau of Land Management. (2012, April). *Pryor Mountain Wild Horse Range quick facts.* Retrieved from http://www.blm.gov/pgdata/etc/medialib/blm/mt/field_ offices/billings/wild_horses/popular_pages.Par.94314.File.dat/FinalPMWHR_Quick_Facts.pdf

U.S. Bureau of Land Management. (2012, October 15). *Pryor Mountain Wild Horse Range.* Retrieved from http://www.blm.gov/mt/st/en/fo/billings_field_office/ wildhorses.html

U.S. Bureau of Land Management. (2013, March 8). *Wild horse and burro quick facts.* Retrieved from http://www.blm.gov/wo/st/en/prog/whbprogram/history_and_facts/quick_facts.html

U.S. Bureau of Land Management, Billings Field Office. (2001). *Environmental assessment and gather plan, Pryor Mountain Wild Horse Range, FY2001 wild horse gather and selective removal* (EA #MT-010-1-44). Billings, MT: Bureau of Land Management, Billings Field Office.

U.S. Bureau of Land Management, Billings Field Office. (2003, April). *Environmental assessment, Pryor Mountain Wild Horse Range, FY03 fertility control on select young wild horse mares; selective removal of young wild horse stallions* (EA #MT-010-03-14). Billings, MT: Bureau of Land Management, Billings Field Office.

U.S. Bureau of Land Management, Billings Field Office. (2009, May). *Pryor Mountain Wild Horse Range/Territory environmental assessment MT-010-08-24 and herd management area plan.* Billings, MT: Bureau of Land Management, Billings Field Office. Retrieved from http://www.blm.gov/pgdata/etc/medialib/blm/mt/field_offices/billings/wild_horses.Par.30079.File.dat/pmwhrFINAL.pdf

U.S. Department of Agriculture, Centers for Epidemiology and Animal Health. (2006, September). *Equine infectious anemia (EIA)* (APHIS Info Sheet). Washington, DC: U.S. Department of Agriculture, Animal and Plant Health Inspection Service. Retrieved from http://www.aphis.usda.gov/vs/nahss/equine/eia/eia_info_sheet.pdf

U.S. National Park Service. (2006). *Management policies 2006.* Washington, DC: Government Printing Office.

van Dierendonck, M., & Goodwin, D. (2005). Social contact in horses: Implications for human-horse interactions. In F. de Jonge & R. van den Bos (Eds.), *The human-animal relationship: Forever and a day* (pp. 65–82). Assen, Netherlands: Royal Van Gorcum BV.

Vavra, M. (2005). Livestock grazing and wildlife: Developing compatibilities. *Rangeland Ecology & Management, 58*(2), 128–134. doi: 10.2111/1551-5028(2005)58<128:LGAWDC>2.0.CO;2

Warshaw, M. (2010, August). Dredge-spoil islands—Town Marsh and Carrot Island. *Beaufort, North Carolina History.* Retrieved from http://beaufortartist.blogspot.com/2010/08/dredging-created-islands.htmlWild horses of North Carolina. (1902, December 21). *New York Times*, p. 27.

Warshaw, M. (2012, April). Carrot Island. *Beaufort Harbor* [Web log]. Retrieved from http://beaufortinlet.blogspot.com/2012/04/1854-map-carrot-island.html

Willis, M. (1999). Shackleford Banks, NC, wild horses free of EIA: Roundup on Shackleford Banks, January 16–22, 1999. *Caution: Horses, 4*(3). Retrieved from http://asci.uvm.edu/equine/law/articles/shackle.htm

Wiss, Janney, Elstner Associates & John Milner Associates. (2007). *Portsmouth Village cultural landscape report.* Atlanta, GA: U.S. National Park Service, Southeast Regional Office. Retrieved from http://www.nps.gov/calo/parkmgmt/upload/CALO Portsmouth Village CLR_Site History.pdf

Wood, C.H. (2002). *Body condition scoring for your horse* (University of Maine Cooperative Extension Publications, Bulletin 1010). Retrieved from http://umaine.edu/publications/1010e/

Wood, G.W., Mengak, M.T., & Murphy, M. (1987). Ecological importance of feral ungulates at Shackleford Banks, North Carolina. *American Midland Naturalist, 118*(2), 236–244.

Zedler, J.B., & Callaway, J.C. (2001). Tidal wetland functioning. *Journal of Coastal Research*, Special Issue 27, 38–64.

Acknowledgments

About 20 years ago, I began to research my first book, *Hoofprints in the Sand: Wild Horses of the Atlantic Coast.* The experience enriched my life in many ways. I learned a tremendous amount about how the natural behavior of these horses differs from that of their domestic counterparts. Following herds through marsh and dune, I gained a greater appreciation of the complexity of their apparently simple lives in the wild. Similarly, with the writing of *Wild Horse Dilemma*, many helpful and extraordinary people have earned my gratitude, and many have become friends.

Dr. Jay F. Kirkpatrick, senior scientist at the Science and Conservation Center at ZooMontana in Billings and author of *Into the Wind: Wild Horses of North America*, has devoted much of his career to preserving the health, rights, and dignity of wild horses. He developed and implemented the immunocontraceptive program in use with many species of wildlife, including horses. He answered my questions, forwarded useful documents, crafted an insightful preface, and generously reviewed the manuscript before publication.

Don Höglund, DVM, is an internationally esteemed leader in horse training and management and the author of *Nobody's Horses*, a riveting book about the rescue of wild horses from the White Sands Missile Range. He has implemented numerous large-scale equine programs, including the Department of the Interior's Wild Horse Prison Inmate Training Program, which teaches prisoners to gentle horses while providing training for adoptable mustangs. His love and admiration for horses is evident in all that he does. When I approached him with a few questions, he responded enthusiastically and sent me a number of articles that shaped the backbone of my manuscript. I am grateful for his support and encouragement along the way and fortunate that he agreed to review the manuscript.

Dr. E. Gus Cothran, Texas A&M University's renowned expert on the genetics of wild and domestic horses, helped me to understand the significance of the Q-ac variant in certain wild horse herds and the concept of minimum viable population. He also found time in his busy schedule to review the manuscript. His research is the cornerstone of wild horse management, and I have cited it extensively.

Dr. Sue Stuska, the wildlife biologist at Cape Lookout National Seashore who oversees the Shackleford Banks herd, has corresponded regularly about the status of the horses. When I visited, she taught me how to identify individuals, and showed me her dynamic census chart that tracks the members of each band and where they were last sighted. She explained how the current management plan makes optimal use of the existing gene pool by monitoring family lineage and contracepting certain mares. Sue also gave of her valuable time to review my manuscript for accuracy.

Doug Hoffman, wildlife biologist at Cumberland Island National Seashore, helped me to understand the Park Service perspective on the horses living there and corrected my assumptions and misinformation. He generously drove me to key parts of the island that I could not otherwise reach, and my time with him was the highlight of the trip.

Karen McCalpin, director of the nonprofit Corolla Wild Horse Fund, Inc., found time in her impossibly busy schedule to meet me and discuss the genetic crisis facing the herd. She also reviewed the manuscript. Karen and the other members of the organization—mostly volunteers—have upended their lives to secure protection for these horses. Karen produced a beautifully written blog highlighting the triumphs and tragedies of the herd. It can be accessed at www.corollawildhorses.com

Dr. Ronald Keiper, Distinguished Professor of Biology (emeritus) at Penn State University, was one of the first scientists to study the behavior of horses in the wild, answered my questions about the foaling rate of lactating mares and shared his groundbreaking research detailing the behavior of the Assateague horses.

Wesley Stallings, former manager of the Corolla herd, took me in his truck several times as he patrolled the Outer Banks north of Corolla, following the movements of wild bands and logging herd data in his notebook. Sometimes we climbed on the roof of the truck to scout for horses. Sometimes Wes climbed a tree for a better view. At one point we encountered a flooded hollow and

were forced to don hip boots and slog through surging currents occupied by cottonmouth snakes to evaluate the health of a newborn colt. I am grateful he allowed me to participate in his daily adventures.

Steve Edwards, by day an attorney for Isle of Wight County, Virginia, works magic in rehabilitating injured Corolla and Shackleford horses. At his farm, Mill Swamp Indian Horses in Smithfield, Va., he teaches children how to train wild horses with natural horsemanship techniques. He also established an off-site breeding program to preserve the herd's rare genes in case of disaster in the wild. Steve has been extremely supportive and helpful throughout the writing of this book, and has brought his expertise to the task of reviewing this book for accuracy before publication.

In 2012, I visited Mill Swamp and was captivated by the sight of children working in the round pen with young horses, many of them recently brought in from the wild. To this point, I had great esteem for the wild horses living on North Carolina's Outer Banks, but had never ridden one. I found these Colonial Spanish Horses astonishingly surefooted, brave, rugged, and smooth-gaited. The climax of my visit was a ride through the inky forest astride Manteo, a wild-born black stallion. He never missed a step despite exposed roots, steep embankments, deep pools, and deer crashing gracelessly through the underbrush. For the better part of an hour, we trotted and cantered through darkness so complete, I could not see my hand in front of my face. I had recently been injured in a riding accident, and I was working through many horse-related fears. It was a profound and humbling

experience to trust a once-wild stallion to find his way through darkness that left me blind.

D. Phillip Sponenberg, DVM, PhD, helped me to understand the genetic underpinnings of coat color and its implications for the free-roaming Banker horses. In his review of the manuscript, he offered great insights on Spanish horse origins and genetics, and his comprehensive articles on that topic were a valuable resource.

Carolyn Mason, president of the Foundation for Shackleford Horses, Inc., accompanied me to Shackleford Banks and generously shared her wealth of knowledge about the horses. She introduced me to the Banker Horses grazing in her yard, gentled animals awaiting adoption. My heart melted when a young gelding named Adagio followed me like a puppy and courted hugs and scratching.

Woody and Nena Hancock loaded me and my cameras into their boat and searched island and marsh for members of the Cedar Island herd. They introduced me to Bucky, the most genetically valuable horse on Cedar Island, and her 2-week-old look-alike filly, Gay; a mare who prefers the company of three burly wild bulls; and Shack, the robust sorrel patriarch whose photograph graces the front cover of *Wild Horse Dilemma*. It was a profound, almost holy experience to stand calf-deep in warm estuarine waters under a moody sky, surrounded by peaceful wild horses, splashing pelicans, and wind-licked marsh grass.

Laura Michaels, the Park Service ranger in charge of pony care, took me behind the scenes to meet the Ocracoke horses. I also met Wenzel, Doran, Sacajawea, and Jitterbug, the Shackleford horses who will revitalize the Ocracoke herd. I even scratched the neck of the lovely black-and-white mare Easter Lady after admiring her from afar for years.

Roe Terry, former public relations specialist of the Chincoteague Volunteer Fire Company, invited me to the workshop where he carves graceful wooden waterfowl and discussed the challenges faced by the hardworking firefighters. Besides managing the herd of free-roaming ponies, these dedicated people donate their time to provide tax-free fire suppression, search and rescue, and emergency medical services in a town of 4,400 permanent residents that receives roughly 1.5 million visitors a year. He also granted me access to the optimal vantage point for the world-famous Chincoteague Pony Swim: whereas most onlookers stood in a field behind an orange fence, out of harm's way, I was able to stand directly on the grassy landing where the horses regained solid ground after swimming the channel from Chincoteague National Wildlife Refuge. Ponies rose out of the water like mythical creatures of the sea, dripping wet and looking very pleased with themselves. My feet were in their hoofprints, and occasionally I dove for cover as a stallion thundered by in pursuit of a rival. It was a magical experience.

Denise Bowden, his successor at the fire company, cheerfully supplied me with useful information. Her passion for the horses and the refuge are evident, and her enthusiasm enhances the overall festivity of Pony Penning week.

Lou Hinds, former manager of the Chincoteague NWR, took me around the refuge to show me unequivocal signs of dramatic environmental change, such as tree trunks, light poles, and chunks of peat that had once been on the bay side of Assateague until island migration situated them squarely on the beach. Studying the dynamic nature of barrier islands is one thing; seeing the evidence of their migration is another thing entirely.

Pam Emge, co-author of *Chincoteague Ponies: Untold Tails,* can identify all of the Chincoteague wild ponies and knows the intimate details of their relationships and lineages. She reviewed part of the manuscript, corrected errors, and filled in details.

Anthropologist Karen Dalke of the University of Wisconsin-Green Bay shared her doctoral dissertation and other writings, which provide unique perspective on what we feel about wild horses and how we define them. She also reviewed the manuscript prior to publication.

Paula Gillikin, manager of the Rachel Carson North Carolina National Estuarine Research Reserve, assisted me in researching the horses of Carrot Island and vicinity.

Philip Howard, nephew of Marvin Howard (1897–1969), who led Ocracoke's mounted Boy Scout troop in the late 1950s, and grandson of the legendary horseman Homer Howard, allowed me to use excerpts from his Web site detailing his family history with the wild ponies.

Allison Turner, biological science technician at Assateague Island National Seashore, supplied excellent information about the Maryland herd and shared Park Service photographs that vividly show the bites and kicks that occur when people get too close to wild horses.

DeAnna Locke, administrator of the Ocracoke Preservation Society, let me pore over and digitize the organization's fascinating scrapbooks, which included many pictures of the island's mounted Boy Scout troop.

Tim Ferry and Flickr user rich701 allowed me to reprint some of their unique historical photographs of the Chincoteague herd.

Craig Downer, a wild-horse ecologist and activist on the board of directors of the Cloud Foundation, shared several of his writings with me on the subject of mustang management.

Jean Bonde of the Buy-Back Babes has a contagious enthusiasm for the Chincoteague Ponies and many tales to tell. These ponies have a large cult following, and her e-mail group recounted the details of their lives—celebrating the romance of Copper Moose and Scotty's ET, pondering Rip Tide's status within Surfer Dude's band, and speculating on why Queenie and Suzy Sweetheart were wandering around the wildlife loop.

Tabetha Fenton of Barefoot Minis helped with proofreading and offered enthusiastic support.

Special thanks to my mother, Joyce Urquhart, who is an exceptionally good proofreader. She read every word of the manuscript and discovered errors that other readers missed.

My husband, Alex, is committed to giving me space in which to create and assisting me wherever possible in my creative pursuits. Besides proof-reading my manuscripts, he is the behind-the-scenes man who maintains the household, runs to the post office, and brings in the bird feeders at night so the bears don't destroy them. He is the love of my life and I am thankful every day that we are together.

About the Author

Bonnie Urquhart Gruenberg is a multifaceted person who wishes that sleep were optional. She is the author of the award-winning textbook *Birth Emergency Skills Training* (Birth Guru/Birth Muse, 2008); *Essentials of Prehospital Maternity Care* (Prentice Hall, 2005); and *Hoofprints in the Sand: Wild Horses of the Atlantic Coast* (as Bonnie S. Urquhart; Eclipse, 2002), as well as articles in publications as dissimilar as *Equus* and the *American Journal of Nursing.* She is an artist and photographer and has illustrated all her own books.

By profession, she is a Certified Nurse-Midwife and Women's Health Nurse Practitioner who welcomes babies into the world at a freestanding birth center in Lancaster County, Pa. She obtained her MSN from the University of Pennsylvania after completing her BSN at Southern Vermont College, and she spent 10 years attending births in tertiary-care hospitals before returning to out-of-hospital practice. Prior to her career in obstetrics, she worked as an urban paramedic in Connecticut.

Horses have been her passion from infancy. For nearly two decades, she has spent countless hours researching and photographing the private lives of wild horses in both Western and Eastern habitats. She has been riding, training, teaching, and learning since her early teens, from rehabilitating hard-luck horses to wrangling trail rides in Vermont and Connecticut. In her vanishing spare time, she explores the hills and hollows of Lancaster County astride her horses Andante and Sonata.

More information and a collection of her photographs can be found at her Web site, www.BonnieGruenberg.com Additional information about the Atlantic Coast horse herds is on the Web at www.WildHorse Islands.com

If you liked this book, you may enjoy other titles by the author:

The Wild Horse Dilemma: Conflicts and Controversies of the Atlantic Coast Herds (Quagga Press, 2015)
The Hoofprints Guide Series (Quagga Press, 2015)
Assateague
Chincoteague
Corolla
Ocracoke
Cumberland Island

Forthcoming

Wild Horse Vacations: Your Guide to the Atlantic Wild Horse Trail with Local Attractions and Amenities (Quagga Press, 2015)
Wild Horses! A Kids' Guide to the East Coast Herds (Quagga Press, 2015)

Visit QuaggaPress.com for details.

CPSIA information can be obtained
at www.ICGtesting.com
Printed in the USA
LVOW04s1735050716
495171LV00018B/973/P